Inorganic-Whisker-Reinforced Polymer Composites

Synthesis, Properties and Applications

T0187699

Inorganic-Whisker-Reinforced Polymer Composites

Synthesis, Properties and Applications

Qiuju Sun • Wu Li

Science Press
Beijing

CRC Press
Taylor & Francis Group
Boca Raton London New York

CRC Press is an imprint of the
Taylor & Francis Group, an **informa** business

CRC Press
Taylor & Francis Group
6000 Broken Sound Parkway NW, Suite 300
Boca Raton, FL 33487-2742

First issued in paperback 2019

ISBN-13: 978-1-4987-0067-2 (hbk)
ISBN-13: 978-0-367-37747-2 (pbk)

Library of Congress Cataloging-in-Publication Data

Sun, Qiuju.
   Inorganic-whisker-reinforced polymer composites : synthesis, properties, and applications / Qiuju Sun and Wu Li.
      pages cm
   Includes bibliographical references and index.
   ISBN 978-1-4987-0067-2
   1. Composite materials. 2. Fibrous composites. 3. Polymers. I. Li, Wu, 1958- II. Title.

TA418.9.C6S793 2016
541'.2254--dc23   2015016371

Visit the Taylor & Francis Web site at
http://www.taylorandfrancis.com

and the CRC Press Web site at
http://www.crcpress.com

Contents

Preface

Inorganic whiskers are acicular single-crystal fiber materials with certain length-to-diameter ratios and excellent physico-chemical properties and good mechanical properties, such as high strength, high hardness, and high heat resistance. Filled into polymers, inorganic whiskers can improve the strength, toughness, wear resistance, heat resistance, and processing performance of polymers, and therefore have become a new type of filling modification material. This book first discusses the importance of the modification of polymer materials, followed by a description of the variety, characteristics, surface treatment, and evaluation methods of inorganic whiskers in the market. It then summarizes the preparation methods and performance analysis of polymers filled with inorganic whiskers. Finally, based on the authors' years of study of calcium carbonate whiskers as fillers, this book introduces the surface treatment methods for and factors influencing calcium carbonate whiskers. Combined with a consideration of research on polypropylene filled with calcium carbonate whiskers, this book systematically and comprehensively discusses the main preparation methods and possible problems and solutions, and summarizes the latest applications and research progress on polymers filled with inorganic whiskers in China and elsewhere.

This book will be useful for teachers, researchers, and master's and PhD students in the fields of polymer materials and engineering at universities and research institutes. It also provides a reference for technology developers in the relevant fields in the industry.

First and foremost, we express our deepest gratitude to Shenyang Normal University, Qinghai Institute of Salt Lakes (Chinese Academy of Sciences), and China Science Press for helping us publish this book. Likewise, we extend our thanks to Deyan Zhao of Shenyang Normal University, Yanfeng Cui, Donghai Zhu, Chengcai Zhu, Yabin Wang, Li Dang, Yanwei Jing, Yaoling Zhang of Qinghai Institute of Salt Lakes, and Qing Meng, who have offered great help in the translation of this book.

Qiuju Sun
Wu Li

Chapter 1

Introduction

A substance that can meet specific morphological and physical performance requirements is referred to as a material. Materials are vital to modern industrial and high-tech advances, constituting an important basis for human survival and development. Materials can be divided according to their chemical composition into metallic, inorganic nonmetallic, and organic polymer materials. Metallic materials are composed of metal atoms such as iron, copper, aluminum, and alloy steel. Inorganic nonmetallic materials are composed of inorganic compounds such as glass, ceramics, and cement. Organic polymer materials are composed mainly of two elements—carbon and hydrogen—with C–C covalent bonds forming their basic structure; examples of such compounds include cottons, linens, silks, plastics, rubber, synthetic fibers, and others.

As a rising star in the field of materials, polymer materials have widespread applications because of their unique features, such as availability of raw materials, low cost, low density, favorable optical activity, diversity, high mechanical strength, chemical resistance, and excellent insulation properties. Furthermore, polymer materials can also adapt to various needs, have good processibility, and are suitable for automatic production, and have therefore become an essential material in our everyday lives.

In addition, with the development of the materials industry and improvements in technology, composite materials made from two or more constituent materials with different physical or chemical properties that are combined with appropriate methods and provide the compound effects of the individual component have also become a large species in the materials field. Changxu Shi, who edited the *Material Dictionary*, defined composite materials as new materials that are made of organic polymers, inorganic nonmetallic or metallic, and other types of different materials through composite technology. They retain the main features of the original components of the materials, but also through their combination they acquire special characteristics that the raw materials individually do not possess. In modern materials science, composite materials generally include the materials of fiber reinforced, sheet enhancement, particle reinforced, or self-reinforcing particles. It can be clearly seen that composite materials have extensive meanings, but in industry, composite materials generally refer to high-performance materials, produced by reinforced materials with high strength and high modulus and matrix materials with low-modulus ductility. Among these reinforced materials, one of the most widely used is fibrous material, which has the best performance. Hence when people speak of composite materials they usually refer to fiber-reinforced composite materials, which actually are, in the narrow sense of the term, composite materials.[1]

A composite material is a multiphase system composed of matrix material and dispersed material. The matrix material is usually a continuous phase, which consolidates the dispersion material such as fiber or particles. The composite material is commonly categorized into the class of materials that the matrix material belongs to: metal matrix composites, inorganic nonmetal matrix composites, and polymer-based composites. For example, the metal matrix composite material is a material included in the scope of metallic materials, while the polymer matrix composite material is included in the scope of the polymer material. The inorganic whiskers filled and modified polymers studied in this book belong to the research area of polymer matrix composites and polymer materials.

1.1 Basic Theory of Polymer Materials

1.1.1 Concept of the Polymer Compound

A polymer compound, also known as high molecular weight polymer, is a macromolecule composed of one or more structural units that are connected by covalent bonds.[2,3] The relative molecular mass of the polymer compound (hereinafter referred to as molecular weight) is generally 10,000 Daltons or more.

For example, polypropylene is made from propylene structural units that are connected to each other.

$$\cdots - CH_2 - \underset{\underset{CH_3}{|}}{CH} - CH_2 - \underset{\underset{CH_3}{|}}{CH} - CH_2 - \underset{\underset{CH_3}{|}}{CH} - \cdots$$

For convenience, they are often abbreviated as shown in Figure 1.1a.

The end groups are often omitted because of their small size. Among them, Figure 1.1b shows a repeating unit and the brackets indicate repeat connection; n represents the number of repeat units, also known as the degree of polymerization. It is an index of the molecular weight. A small molecular weight compound, which forms the structural unit, is called a monomer. In other words, polypropylene is made from a propylene monomer by a polyaddition reaction (addition polymerization reaction). The polymerization reaction is shown as

$$n CH_2 = \underset{\underset{CH_3}{|}}{CH} \longrightarrow \left[CH_2 - \underset{\underset{CH_3}{|}}{CH} \right]_n$$

A polymer with only one type of monomer is referred to as a homopolymer, such as polypropylene, poly(vinyl chloride), poly(methyl methacrylate), and so on. A polymer with two or more types of monomers is called a copolymer. For example, ethylene–propylene copolymer is synthesized via ethylene and

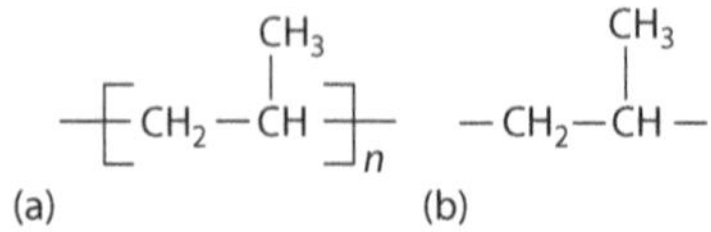

$$\left[\!\!\begin{array}{c} CH_3 \\ | \\ CH_2-CH \end{array}\!\!\right]_n \qquad \begin{array}{c} CH_3 \\ | \\ -CH_2-CH- \end{array}$$

(a)   (b)

Figure 1.1   Polypropylene units. (a) Abbreviated unit. (b) Repeating unit.

propylene copolymerization, and vinyl chloride–vinyl acetate copolymer is produced using vinyl chloride and vinyl acetate copolymerization. With containing two types of structural units (indicated as A and B respectively) copolymer as an example, the arrangements of the two structural units in the macromolecular chain are of the following four types:

1. Random copolymers with two structural units arranged in no particular order

   ----– ABABBBABAABBAAA –----

2. Alternating copolymers with regular alternation of two structural units

   ----–ABABABABABABABAB –----

3. Block copolymers with two or more homopolymer sub-units, such as AB type, ABA type, and so forth

   AB type: ----–$(AAAAAAA)_x$——$(BBBBBBB)_y$ ---

   ABA type: ---–$(AAAAAAA)_x$—$(BBBBBBB)_y$—$(AAAAAAA)_z$---

4. Graft copolymers with one structural unit forming a long chain as the main chain, and another structural unit forming the side chains that are connected with the main chain

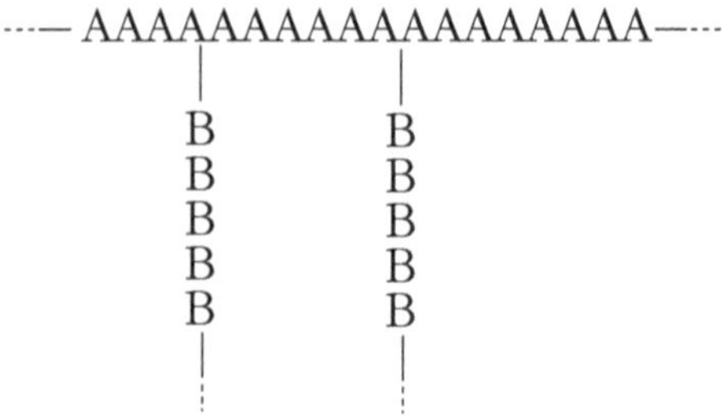

   ----– AAAAAAAAAAAAAAAAAAAA –----
            | |
            B   B
            B   B
            B   B
            B   B
            B   B
            | |
            ⋮   ⋮

Copolymers with different arrangements of the structural units have different performances.

Polymers such as poly(ethylene terephthalate) and poly-hexamephylene adipamide have other features. For example, the structure of poly(ethylene terephthalate) is

$$\left[\!\!\begin{array}{c} \overset{O}{\underset{\|}{C}}-\!\!\bigcirc\!\!-\overset{O}{\underset{\|}{C}}-O-CH_2CH_2O \end{array}\!\!\right]_n$$

The repeating unit is composed of two structural units:

$$-\overset{O}{\underset{\|}{C}}-\!\!\bigcirc\!\!-\overset{O}{\underset{\|}{C}}- \text{ and } -OCH_2CH_2O-$$

These two structural units have fewer atoms than the corresponding monomeric (terephthalic acid and ethylene glycol) because these two monomers lose water molecules in the polymerization reaction (condensation polymerization). The polymerization reaction is

$$n\text{HOOC}-\!\!\bigcirc\!\!-\text{COOH} + n\text{HOCH}_2\text{CH}_2\text{OH} \longrightarrow \left[\!\!\begin{array}{c}\overset{O}{\underset{\|}{C}}-\!\!\bigcirc\!\!-\overset{O}{\underset{\|}{C}}-OCH_2CH_2O\end{array}\!\!\right]_n + (2n-1)H_2O$$

Polymers are primarily utilized as materials processing unique mechanical strength, which is closely related to the molecular weight of a polymer. A schematic relationship between the mechanical strength of a polymer and the molecular weight is shown in Figure 1.2.

Point A has the lowest molecular weight; above point A, the mechanical strength increases rapidly with molecular weight; above the critical point B, mechanical strength increases slowly, while after point C, there is no significant increase in strength.

The low molecular compound generally has a fixed molecular weight, while the polymer is a homogeneous mixture of macromolecules with different molecular weights. Therefore, the molecular weight and the degree of polymerization are average values. According to different statistical methods, the polymers mainly have three average molecular weights: number-average molecular weight $(\overline{M_n})$, weight-average molecular weight $(\overline{M_w})$, and viscosity-average molecular weight

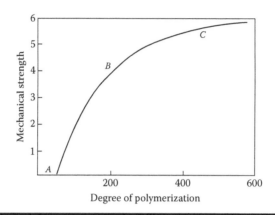

Figure 1.2   The relationship between the mechanical strength and molecular weight of a polymer.

$\left(\overline{M_{\eta}}\right)$. The order of these three types of molecular weight is $\overline{M_{w}} > \overline{M_{\eta}} > \overline{M_{n}}$. The distributions of the polymer molecular weight are indicated by polydispersity. The polydispersity of the molecular weight of the polymer can be represented by a distribution curve or distribution index. Figure 1.3 is a typical distribution curve of the molecular weight. The relative sizes of the $\overline{M_{w}}$, $\overline{M_{\eta}}$, and $\overline{M_{n}}$ are indicated on the abscissa.

The distribution index is defined as the ratio of the weight-average molecular weight to the number-average molecular

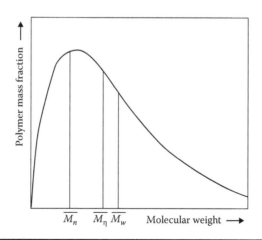

Figure 1.3   Typical molecular weight distribution curve.

weight, that is, $\overline{M_w}/\overline{M_n}$. For a system with homogeneous molecular weight, $\overline{M_w} = \overline{M_n}$, that is, $\overline{M_w}/\overline{M_n} = 1$. The larger the $\overline{M_w}/\overline{M_n}$ ratio is, the wider distribution of the molecular weight.

In addition to the average molecular weight, the distribution of the molecular weight is an important factor that affects the properties of the polymer. A low molecular weight macromolecule decreases the strength of the polymer, whereas a high molecular weight macromolecule makes plasticization and molding difficult. Therefore, different polymer materials should have an appropriate molecular weight distribution.

1.1.2 Nomenclature of Polymer Compounds

Polymer compounds can be named according to the systematic nomenclature or the traditional nomenclature.

1.1.2.1 Systematic Nomenclature

The systematic nomenclature is also called the International Union of Pure and Applied Chemistry (IUPAC) nomenclature, which usually includes the following steps:

1. Identify the repeating unit of the structure.
2. Sort the order of each atom of the repeating unit according to the priority of the organic group.
3. Name the repeating unit and add parentheses according to the nomenclature rule of small molecular weight organic compound.
4. Add the prefix "poly" before the names.

Because of the elaborateness of the systematic nomenclature, it is usually used to name new polymers or for academic communications, as shown in Table 1.1.

Table 1.1   The Repeating Units and Nomenclature of Some Polymers

Traditional Nomenclature	Repeating Unit	Systematic Nomenclature
Polyethylene	$-CH_2-CH_2-$	Polyethylene
Polypropylene	$-CH-CH_2-$ $\qquad$ | $\qquad CH_3$	Poly (1-methylethylene)
Polystyrene	$-CH-CH_2-$ $\qquad$ ⌬	Poly (1-phenylethylene)
Poly(methyl methacrylate)	$\qquad CH_3$ $\qquad$ | $-C-CH_2-$ $\qquad$ | $\qquad COOCH_3$	Poly [1-(methoxycarbonyl)-1-methyl ethylene]
Poly(hexamethylene adipamide)	$-NH(CH_2)_6NHCO(CH_2)_4CO-$	Poly (iminohexamethylene imino adipoyl)

1.1.2.2 Traditional Nomenclature

The traditional nomenclature is commonly used. Polymers are named according to one of the following:

1. *By the monomer's or hypothetical monomer's name, preceded by the word "poly."* For example, the polymer obtained by the polymerization of ethylene is called polyethylene. The polymer obtained by the polymerization of propylene is called polypropylene. Others such as polyvinyl chloride, polystyrene, and poly(methyl methacrylate) are the polymers of vinyl chloride, styrene, and methyl methacrylate, respectively. The reaction of polyethylene and poly(vinyl chloride) is

$$nCH_2{=}CH_2 \longrightarrow \left[CH_2-CH_2 \right]_n$$

$\qquad$ Ethylene $\qquad\qquad\qquad$ Polyethylene

$$n\text{CH}_2\text{=CH} \longrightarrow \left[\text{CH}_2 - \text{CH}\right]_n$$

Cl	Cl
Chloroethylene	Polyvinyl chloride

As a special case, polyvinyl alcohol is not obtained by the polymerization of ethylene alcohol, because ethylene alcohol has an unstable enol structure that can easily form aldehyde by isomerism; therefore, vinyl alcohol is only a hypothetical monomer of poly(vinyl alcohol). For the majority of polymers prepared by means of a double bond addition reaction of alkene and diene monomers, this nomenclature method is simple, intuitive, and widely used.

For polymers obtained by two or more kinds of alkene monomers by the addition reaction, a dash is always added between two monomers and the prefix "poly" is affixed to the word: "monomer name 1 + monomer name 2 + copolymer." For example, the copolymer of methyl methacrylate and styrene is named "poly(methyl methacrylate–styrene)" or "methyl methacrylate–styrene copolymer"; the polymer that is obtained by copolymerization of acrylonitrile, butadiene, and styrene is named "poly(acrylonitrile–butadiene–styrene)" or "acrylonitrile–butadiene–styrene copolymer."

In the international nomenclature, "-co-, -alt-, -b-, -g-" are often inserted between two monomers to represent random copolymerization, alternating copolymerization, block copolymerization, and graft copolymerization, respectively. In random copolymer names, the former is the main monomer, and the latter is the secondary monomer. In block copolymer names, the order of monomers represents the order of polymerization, whereas in graft copolymer names, the former is the main chain and the latter is the branched chain.

A polymer obtained through a polycondensation reaction of two kinds of monomers containing different functional groups cannot be named with this method, and should be named according to the chemical structure. For example, the polymer obtained by polycondensation of terephthalic acid and ethylene glycol, containing the characteristic group in Scheme 1.1a is named "poly(hexamephylene adipamine)"; the polymer formed from hexamethylene diamine and adipic acid, containing functional groups in Scheme 1.1b, is named "poly(hexamephylene adipamide)." The reaction equation is

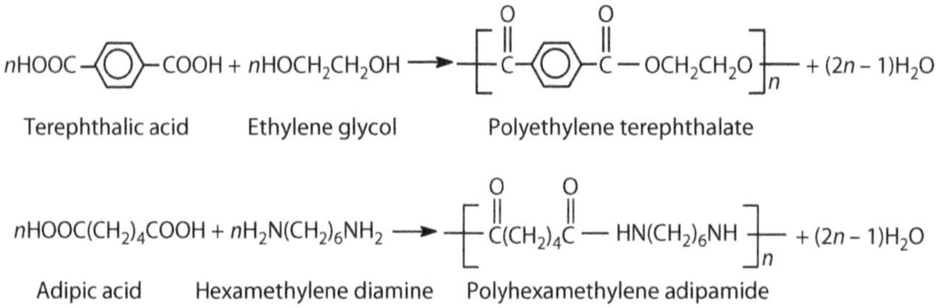

$$n\text{HOOC} \text{—} \langle O \rangle \text{—COOH} + n\text{HOCH}_2\text{CH}_2\text{OH} \longrightarrow \left[\overset{O}{\overset{\|}{C}} \text{—} \langle O \rangle \text{—} \overset{O}{\overset{\|}{C}} \text{—OCH}_2\text{CH}_2\text{O} \right]_n + (2n-1)\text{H}_2\text{O}$$

Terephthalic acid   Ethylene glycol   Polyethylene terephthalate

$$n\text{HOOC}(\text{CH}_2)_4\text{COOH} + n\text{H}_2\text{N}(\text{CH}_2)_6\text{NH}_2 \longrightarrow \left[\overset{O}{\overset{\|}{C}}(\text{CH}_2)_4\overset{O}{\overset{\|}{C}} \text{—HN}(\text{CH}_2)_6\text{NH} \right]_n + (2n-1)\text{H}_2\text{O}$$

Adipic acid   Hexamethylene diamine   Polyhexamethylene adipamide

2. *By functional groups.* Polymers are named by characteristic groups in the main chain. For example, the polymers containing Scheme 1.1c in their molecules are collectively referred to as "polyamide"; the polymers containing Scheme 1.1d in their molecules are collectively referred to as "polyester"; the polymers containing Scheme 1.1e in their molecules are collectively referred to as "polyurethane"; the polymers containing –o– in their molecules go by the general name of "polyether."

3. *By the composition of the polymer.* This nomenclature is commonly used for thermosetting resin and rubber-like polymers. In this method, the word "resin" or "rubber" is

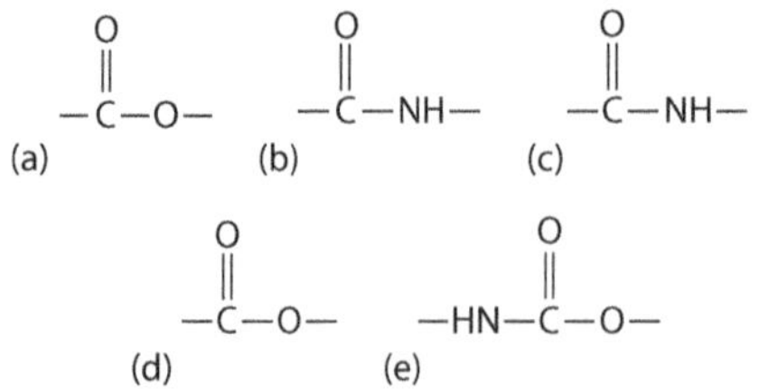

$$\overset{O}{\overset{\|}{—C—O—}} \qquad \overset{O}{\overset{\|}{—C—NH—}} \qquad \overset{O}{\overset{\|}{—C—NH—}}$$
(a)   (b)   (c)

$$\overset{O}{\overset{\|}{—C—O—}} \qquad \overset{O}{\overset{\|}{—HN—C—O—}}$$
(d)   (e)

Scheme 1.1

added at the end of the monomer name or abbreviation. For example, chloroprene rubber is formed by the polymerization of chloroprene; phenolic resin is formed by the polymerization of phenol and formaldehyde; epoxy resin is formed by the polymerization of epoxy compounds; and styrene butadiene rubber is made by the copolymerization of butadiene and styrene.

4. *By trade name or common name.* The trade or proprietary name is assigned by the product manufacturer, and they emphasize merchandise or variety, for example, the trade name of polyamide is "nylon"; other trade names include, for example, Teflon (polytetrafluoroethylene) and Celluloid (nitrocellulose). Names such as "organic glass" (poly(methyl methacrylate)), "Bakelite" (phenolic resin), and "electric jade" (urea formaldehyde resin) have also been widely used.

   Trade names are also widely used in synthetic fibers. For example, polyethylene terephthalate is called Dacron; polypropylene, polypropylene fiber; polyacrylonitrile, Acrylon; poly(vinyl chloride), chloro fiber; and polyamide, nylon. The numbers following the word nylon have different meanings: The first number represents the number of carbon atoms of diamine, and the second number indicates the number of carbon atoms of dicarboxylic acid. If only a single-digit number is used, it indicates the number of carbon atoms in the lactam or amino acids. For example, nylon-66 is a poly(hexamethylene adipamide) compound synthesized from hexamethylenediamine and adipic acid. Nylon-1010 is poly(decamethylene sebacamide) synthesized by decamethylenediamine and sebacic acid. Nylon-6 is polycaprolactam synthesized from ε-caprolactam. Abbreviations of common polymers are shown in Table 1.2.

1.1.3 Classification of Polymer Compounds

Several kinds of classification methods are used for polymer compounds.

Table 1.2   Abbreviations of Common Polymers

Polymer	Abbreviation	Polymer	Abbreviation
Polyethylene	PE	Poly(methyl methacrylate)	PMMA
Polypropylene	PP	Poly(ethylene terephthalate)	PET
Polystyrene	PS	Acryonitrile-butadiene-styrene terpolymer	ABS
Poly(vinyl chloride)	PVC	Epoxy resin	EP
Polyacrylonitrile	PAN	Polycarbonate	PC
Polyamide	PA	Polyurethane	PU

1.1.3.1 According to the Source of the Polymer Compounds

Polymers can be divided into natural, modified, and synthetic polymers. A natural polymer refers to a polymer compound existing in nature. Cotton, silk, starch, protein, wood, natural rubber, and so forth that we usually use in clothing, food, housing, and transport are natural polymer materials. A modified polymer, also called a semisynthetic polymer, is a natural polymer compound treated by a chemical reaction. The world's first man-made polymer material—cellulose nitrate—was made from natural cellulose, such as cotton or cotton cloth, that was treated with concentrated nitric acid and concentrated sulfuric acid. A synthetic polymer is a polymer compound synthesized by small molecular weight compounds through chemical methods. Examples of synthetic polymer materials are plastics such as polyethylene, polypropylene, and polyvinyl chloride and fibers such as polyester, nylon, and other synthetic fibers.

1.1.3.2 *According to the Usage of Polymer Materials*

Polymer materials can be divided into plastic, rubber, fiber, adhesive, coatings, and functional polymer six categories.

Plastic is a material that can be plasticized into certain shapes under certain conditions (temperature, pressure, etc.) and can keep its shape unchanged at room temperature and normal atmosphere pressure. According to their performance after heat treatment, plastics can be divided into thermoplastic and thermosetting plastics. A thermoplastic plastic is generally a linear or branched polymer. It melts when heated and solidifies when cooled, and this kind of behavior can be repeated, so the plastic can be used multiple times. The main varieties are polyethylene, polypropylene, polyvinyl chloride, polystyrene, and acrylonitrile-butadiene-styrene terpolymer. Thermosetting plastic is a space network polymer, which is formed by direct polymerization of monomers or by cross-linking of linear prepolymers. Once the solidification is finished, the polymer cannot be heated back to the plasticizing state. The main varieties are phenolic resin, epoxy resin, amino resin, and unsaturated polyester.

Plastic is an important polymer material with extensive application because of its characteristics, such as light weight, electrical insulation, chemical resistance, and good processibility. The prominent drawbacks of plastics include worse mechanical properties than metal materials, low surface hardness, flammability of most varieties, and poor heat resistance. According to the application range, plastics can be divided into general plastics and engineering plastics. General plastics have high yield, low price, and generally mechanical performance, and are used mainly as nonstructural material, such as polyethylene, polypropylene, polyvinyl chloride, polystyrene, and so on. Engineering plastics are generally referred to as structural materials, which can withstand a wide temperature change range and relatively harsh environmental conditions, and have excellent mechanical properties, good heat- and wear-resistant properties, and good dimensional stability. The main varieties are polyamide, polycarbonate, polyoxymethylene.

In recent years, with the rapid development of science and technology, the performance requirements of polymer materials have become higher and higher and the application areas of engineering plastics continue to expand, with their production increasing year by year. Therefore, the boundary between engineering plastics and general plastics becomes ambiguous and difficult to divide sharply. Some general plastics, such as polypropylene, also can be used as structural materials after modification. Therefore, the emphasis of research on plastics modification is on improving the performance of plastics and increasing their functions.

Rubber is an organic macromolecule with high elasticity, also known as an elastomer. It has excellent elasticity in a wide temperature range (–50°C to 150°C). In addition to the unique high elasticity, rubber also has good mechanical, chemical, and wear resistance, as well as good electrical insulation, so it has become an indispensable material in the national economy.

Rubber can be divided into natural rubber and synthetic rubber on the basis of its source. Natural rubber is prepared from rubber-containing plants in nature, such as a high-elasticity material produced from natural rubber trees. Synthetic rubber is a synthetic macromolecular elastic material prepared by artificial methods. The main varieties of synthetic rubber are styrene–butadiene rubber, butadiene rubber, chloroprene rubber, isoprene rubber, butyl rubber, and ethylene–propylene–diene monomer rubber.

Fiber is a thin material with a length many times larger than its diameter and has certain flexibility. The aspect ratio (length and diameter ratio) of a textile fiber is generally greater than 1000:1. Fiber is also divided into two categories: One is a natural fiber, such as cotton, wool, silk, and hemp, and another kind is a chemical fiber that is made from natural or synthetic polymer compounds by chemical processing. Chemical fibers can be divided into man-made fibers and synthetic fibers according to the source of macromolecular compounds and chemical structure. Man-made fibers use natural macromolecular compounds as raw material that undergoes chemical treatment and mechanical processing; synthetic fibers are made from synthetic polymer fibers. The main varieties of synthetic fibers are polyester (Dacron), polyamide (nylon), acrylon, vinylon, and polypropylene fiber. Plastic, rubber, and fiber are what we usually call "three synthetic materials."

Coatings and adhesives are derived from plastics. A coating is a thin protective and decorative film material covering the surface of an object. An adhesive is a material that glues all kinds of materials closely together. Because these two types of materials are mixture products, they are mainly introduced in fine chemical products. Functional polymers were developed only recently and are a new class of polymeric material with the most developmental potential. A functional polymer usually refers to a polymer material with special properties, such as light, electric, magnetic, heat, and so forth. Examples are conductive polymer, liquid crystal polymer, and biopolymer. The focus on this kind of material is not on the mechanical properties of the polymer, but rather on the specific physical, chemical, and biological functions.

This kind of classification method of polymer materials is not very strict, because the same kind of polymer often can have a variety of applications. For example, polyurethane resin is very wear resistant and therefore can be used to produce plastic runway and skate wheels. After foaming, polyurethane can be used to produce foam plastics with different levels of hardness that are used for making furniture, cushions, and insulation materials. In addition, because it is elastic, polyurethane can replace rubber to make soles of sports shoes. When pulled into wires, it also can be prepared into high-strength and high-elastic Lycra fiber. Polyurethane coating is a kind of wear- and water-resistant coating with high performance and therefore can be used for preparation of floor paint and industrial paint with high strength. The adhesive made from polyurethane has very high strength and is an excellent structural adhesive. Furthermore, because polyurethane has excellent biological and blood compatibility, it also has significant applications as a medical material. The ability to adapt for multipurpose needs is an important reason that polymer materials have attracted much attention.

1.1.3.3 According to the Main Chain Structure of the Polymer

The classification based on the composition of the polymer main chain is relatively strict. In this way, polymers can be divided into carbon chain polymer, heterochain polymer, and element organic polymer. The molecular backbone of a carbon chain polymer is composed entirely of carbon atoms. Most of these polymers are formed through addition polymerization of alkene and diene monomers containing double bonds, such as polyethylene, polypropylene, polyvinyl chloride, etc. The molecular chain of a heterochain polymer contains carbon atoms, as well as oxygen, nitrogen, and sulfur atoms. This kind of polymer molecule typically contains a characteristic group, such as polyester, polyamide, polyurethane, polysulfone, and so forth. The molecular chain of elements organic polymer contains no carbon atoms, and is composed of silicon, boron, aluminum, oxygen, nitrogen, sulfur, phosphorus, and other atoms, with the side group composed of organic groups containing carbon and hydrogen, such as silicone rubber (poly(dimethyl siloxane)) and so forth.

1.1.4 Applications of Polymer Materials

Polymer materials are the foundation of modern technological development. Polymer materials, especially synthetic polymer materials, because of their excellent performance, have played an important role in industry; agriculture; national defense; and transportation, as well as in information, life, and other new technology areas. Statistically, polymer materials account for the most demand for materials. Rubber and plastics account for 70% of the total output of synthetic materials; polymer materials account for about 65% of the total weight of aircraft and about 18% of total vehicle weight. In addition, there would be no modern automobile industry if there were no synthetic rubber for tire manufacture. The development of technology of high-resolution photoresist and plastic resins made large-scale integrated circuits possible, thus contributing to today's computer technology.

The applications of polymer materials in modern lives, especially in clothing, food, housing, and transport are too numerous to mention. It is not an exaggeration to say that we live in the world of polymers. Our toothbrushes and water cups are lightweight and convenient plastics. We can fry eggs for breakfast using cookware that is nonstick because the bottom surface is coated with a layer of polymer material called Teflon (PTFE). When we heat food in a microwave oven, the bowls and plates containing food are made of polypropylene materials. Numerous objects in the kitchen are made of polymer materials, such as basins, seasoning boxes, juice bottles, milk cartons, dishwashing liquid bottles, serving baskets, and film wraps for preserving food.

Many articles of clothing are made from polymer materials. Coats are often made of chemical fiber or polyester wool; trousers are made of high-elasticity Lycra, socks are nylon or ammonia cotton; the soles of leather shoes or sneakers are made of polyurethane. Around the house, plastic steel doors and windows, screens, setting doors, intake and drainage pipelines, sunshades, and so forth are also produced from polymer materials. One can see polymer coating everywhere, such as interior walls, refrigerators, furniture, and so on. Outdoor buildings, automobiles, billboard signs, warning signs, and signal signs are also decorated with coatings. On the road, synthetic rubber tires on bicycles, motorcycles, and cars of different sizes make it possible for people to travel conveniently and quickly.

With the rapid development of the polymer material industry, the production of plastic, rubber, and synthetic fiber also increases year by year. Evidence shows that the plastic yield has exceeded the total yield of structural materials such as wood and cement; synthetic rubber production has exceeded that of natural rubber; and synthetic fiber production in the 1980s already reached two times that of cotton, wool and other natural and man-made fibers.

1.2 Modifications of Polymer

Although polymer materials have excellent performance and are widely used, compared with inorganic materials and metal materials, they also have four common weaknesses: not high enough intensity, poor temperature resistance, flammability, and are easy to degrade.[3,4] For example, when we pour hot water into pure water or mineral water bottles, shrinkage and deformation will occur. If rubber gloves are often used in hot water, they will become sticky, adhesive, and unusable. After long exposure to sunlight, a plastic basin or bucket often becomes faded, brittle, and easily broken. These phenomena reflect the polymer material's poor heat resistance and easy degradation.

On the other hand, with the rapid development of the polymer material industry, the application areas of polymeric materials are also expanding. Some novel requirements for the properties of polymer materials have been proposed. The performance of a single polymer material is difficult to meet the needs of practical application. Thus, the development and application of new materials, complementary properties between different materials, and composite modification will become the emerging trend in the development of new polymer materials.

In general, it is easier to modify a polymer material than synthesize a new polymer and make it ready for application. These modifications can be conducted in general plastics and rubber plants; they can be effective and likely to lead to mass production, and so often can solve many problems in industrial production. Therefore, the modification of polymer materials is gaining more and more attention in industry.

Modification of polymer materials has a very broad meaning. In the modification process, both physical and chemical changes can take place. Because of the perplexing relationship between the structure and properties of polymeric materials, when one method is used to improve a certain kind of performance, other properties may also change. Therefore, in the modification practice, extra multiple impacts of some valuable properties of polymer materials must be prevented, and an overall balance should be sought among the conflicting effects.[3,5]

1.2.1 Classification of Modification of Polymer Materials

1. Based on whether there is chemical reaction during modification, modifications can be divided into two categories: physical modification and chemical modification.

   In physical modification there is no chemical reaction or only a minimal chemical reaction during the whole modification process. The modification principle relies mainly on physical interactions (such as adsorption, coordination, or hydrogen bonding effects) between different components and on the morphological changes of the component itself to achieve the modification objective. It includes filling modification by adding small molecules to the polymer, blending and crosslinking modifications between polymers, morphology control and surface modification of polymers, and so on.

The physical modification method is simple, convenient, fast, economic, and easy to operate, and is the most widely used method for modification.

The new materials combined by two or more kinds of polymers in some way are customarily called polymer blends, or polymeric alloys, and have structures and properties different from those of the original component. The composite materials composed of polymer and inorganic materials are called polymer matrix composites. Inorganic materials, especially inorganic fillers such as calcium carbonate, talcum powder, and mica powder, which use inorganic minerals as raw materials, are widely used because of their low cost and convenient filling. For example, the notched impact strength can increase three times by adding 15 portions of superfine calcium carbonate to polyvinyl chloride (PVC)/acrylonitrile butadiene styrene (ABS) (100/8) blend; adding six portions of $CaCO_3$/polybutyl acrylate (PBA)/polymethyl methacrylate (PMMA) composite particles prepared from nanometer calcium carbonate into 100 portions of PVC can increase PVC notched impact strength by two to three times. As another example, through filling silane coupling agent treated glass fiber and nano-zinc oxide into polypropylene, the mechanical performance, flow properties, crystallization rate, and crystallization temperature of the polypropylene are improved significantly.[6]

Chemical modification refers to the modification process in which chemical reactions occur between the backbone, branched chain, and side chain of a macromolecular chain. The modification principle depends mainly on the structural changes of the main chain, branched chain, or side chain. Chemical modifications include copolymerization between different monomers, grafting reaction of macromolecular chains, crosslinking reaction within macromolecular chains, functional group reactions on the macromolecular chains, and so forth. For example, by aggregating a liquid crystal unit into polysiloxane molecular chains through grafting copolymerization, Zhang[7] synthesized polysiloxane with liquid crystal properties. The chemical reaction is as follows:

$$CH_2=CHCH_2O-\!\!\bigcirc\!\!-COO-\!\!\bigcirc\!\!-OOC-\!\!\bigcirc\!\!-C_4H_9$$

+

$$CH_2=CHCH_2O-\!\!\bigcirc\!\!-COO-\!\!\bigcirc\!\!\bigcirc\!\!-OOC-\!\!\bigcirc\!\!-OCH_3$$

$$\text{Hexachloroplatinic acid} \quad \left| \quad CH_3-\!\underset{\underset{CH_3}{|}}{\overset{\overset{CH_3}{|}}{Si}}\!\!-\!\!\left(\!O-\!\underset{\underset{H}{|}}{\overset{\overset{CH_3}{|}}{Si}}\!\right)_{\!7}\!\!O-\!\underset{\underset{CH_3}{|}}{\overset{\overset{CH_3}{|}}{Si}}\!\!-CH_3 \right.$$

↓

$$CH_3-\!\underset{\underset{CH_3}{|}}{\overset{\overset{CH_3}{|}}{Si}}\!\!-O\!\!\left(\!\underset{\underset{R_1}{|}}{\overset{\overset{CH_3}{|}}{Si}}\!\!-O\!\right)_{\!x}\!\!\left(\!\underset{\underset{R_2}{|}}{\overset{\overset{CH_3}{|}}{Si}}\!\!-O\!\right)_{\!y}\!\!\underset{\underset{CH_3}{|}}{\overset{\overset{CH_3}{|}}{Si}}\!\!-CH_3$$

$$R_2=CH_2CH_2CH_2O-\!\!\bigcirc\!\!-COO-\!\!\bigcirc\!\!-OOC-\!\!\bigcirc\!\!-C_4H_9$$

$$R_1=CH_2CH_2CH_2O-\!\!\bigcirc\!\!-COO-\!\!\bigcirc\!\!\bigcirc\!\!-OOC-\!\!\bigcirc\!\!-OCH_3$$

In addition, by grafting liquid crystal units and ionic groups into the polysiloxane backbone at the same time, liquid crystalline ionomers containing sulfonate groups as the side chains were synthesized. The liquid crystalline ionomer has the performance of liquid crystal and the polarity of an ionomer and can be used as a compatibilizer in polymer blends.[8,9] The reaction equation is:

$$CH_3-\!\underset{\underset{CH_3}{|}}{\overset{\overset{CH_3}{|}}{Si}}\!\!-O\!\!\left(\!\underset{\underset{R_1}{|}}{\overset{\overset{CH_3}{|}}{Si}}\!\!-O\!\right)_{\!x}\!\!\left(\!\underset{\underset{R_2}{|}}{\overset{\overset{CH_3}{|}}{Si}}\!\!-O\!\right)_{\!y}\!\!\underset{\underset{CH_3}{|}}{\overset{\overset{CH_3}{|}}{Si}}\!\!-CH_3$$

$$R_2=(CH_2)_{10}COO-\!\!\bigcirc\!\!-N=N-\!\!\bigcirc\!\!-SO_3H$$

$$R_1=CH_2CH_2CH_2O-\!\!\bigcirc\!\!-COO-\!\!\bigcirc\!\!\bigcirc\!\!-OOC-\!\!\bigcirc\!\!-C_5H_{11}$$

Chemical modification has a durable effect, but it is expensive and difficult to operate, so it is generally finished in a resin synthetic factory and rarely used in a polymer material processing factory. Only a crosslinked modification of polymer material can be conducted while processing.

2. Modifications can be divided into overall modification and surface modification according to whole or partial modification.

Overall modification is the modification that occurs in both the interior and the surface of the polymer material. A feature of this kind of modification is that performance changes uniformly. Modification of polymer materials is mostly overall modification, such as filling modification, blending modification, crosslinking modification, morphology control modification, and so forth that were mentioned previously.

Surface modification refers to the modification that occurs only on the surface of a polymer material without further internal modification. Surface modifications of polymeric materials include surface chemical oxidation, corona surface treatment, surface flame treatment, surface heat treatment, surface plasma treatment, surface metallization processing, ion implantation, and surface grafting polymerization. Because surface modification occurs only on the surface of materials, the performance does not change uniformly.

Compared with overall modification, surface modification has the advantage of low cost and is therefore applicable in situations in which only improvements in appearance are required and internal performance is not important or unnecessary. Common examples are modifications to improve surface gloss; hardness; wear; and anti-static, flame retardant, adhesive, printing, and heat resistant properties. For example, polyurethane foam has the advantages of low density, bubble uniformity, and high-temperature and degradation resistance. Given the advantages of its three-dimensional network structure, polyurethane foam can be used for the preparation of foam metals by using it as a skeleton and silver plating on the surface. Foam metal has a unique structure and special properties of high porosity, low density, high permeability, low thermal conductivity, and sound and vibration absorption ability; it is therefore widely applied in the fields of construction, transportation, petrochemical industry, metallurgy, machinery, electronic products, and communication.[10,11]

3. According to the modification that occurred before or after polymer material molding, modifications can be divided into the categories of polymer synthesis modification, processing modification, and post-processing modification.

   Polymer synthesis modification is the introduction of relevant chemical groups into the polymer structure during the manufacture of raw materials. After the synthesis modification according to the performance or function requirement, the modified material has the desired performance or function. For example, the introduction of an aromatic ring and heterocyclic groups into backbone can improve heat resistance of materials; the introduction of a conjugated electron system into the molecular structure can give the material electrical conductivity; the introduction of hydrophilic, flame retardant, and easy dyeing groups through copolymerization or blending technology can give the polymer material new functions.

   Processing modification is the modification that can give a material the desired performance or function by using blending, compounding, filling, or morphological control during the molding process of the polymer material. This kind of modification includes filling modification, blending modification, crosslinking modification, and morphology control modification of polymer materials.

Post-processing modification is physical or chemical processing on the surface of a material, giving it a special function or performance. Examples include resin finishing, deweighting, surface plasma treatment, surface metal treatment, surface grafting polymerization of fiber materials, surface chemical oxidation treatment, surface corona treatment, surface flame treatment, surface heat treatment, surface plasma treatment, surface metallization processing, ion implantation, and surface grafting polymerization of plastic materials.

1.2.2 Modification Methods for Polymer Materials

Modification of polymer materials can be achieved through a large number of methods, including through chemical modification, blending modification, filling modification, fiber composite reinforcing modification, surface modification, coextrusion compound modification and recycling of polymer materials.

1.2.2.1 Chemical Modification

Chemical modification changes the types of atoms or groups of atoms and their combination in the molecular chains through chemical reactions of polymers. The polymer material itself is a kind of chemical synthetic material, so it can be modified by a chemical method. After chemical modification, the structure of the molecular chain changes; and the polymer material has acquired new properties and its application field is expanded.

Rubber vulcanization crosslinking was an early chemical modification method. Block and graft methods are also widely used in polymer modification. One of the successful examples of a block copolymer is a thermoplastic elastomer. It is a new material that can be processed like plastic and has elasticity like rubber. Among graft copolymers, the most widely used one is the acrylonitrile butadiene and styrene copolymer

(ABS). This material has excellent performance and a relatively low price and is widely used in many fields. In addition, two kinds of polymers in the interpenetrating polymer network (IPN) interpenetrate each other and form a continuous network structure of two phases, as in the chemical modification method. The application of IPN is not yet universal. However, it is expected that it will be adopted.

In addition, through chemical modification, new materials with different structures and properties that cannot be achieved by polyaddition or polycondensation methods can be obtained. The most typical example is the preparation of polyvinyl alcohol and polyvinyl alcohol acetal. The main method involves first that polyvinyl acetate is obtained by polymerization of vinyl acetate; next the ester bonds on the side chain of polyvinyl acetate undergo hydrolysis or alcoholysis, and then polyvinyl alcohol can be obtained. Furthermore, if the hydroxyl and formaldehyde or butyraldehyde on the side chains of polyvinyl alcohol undergo acetalation, polyvinyl butyral or polyvinyl formal can be prepared. Polyvinyl formal after spinning is what we often call vinylon. Vinylon is used mainly for making knitted fabrics such as garment, cotton sweaters and trousers, and jerseys, and also can be used for surgical operation suture, canvas, nets, bicycle tire cord, filtration materials, and so on.

The polymerization reaction of polyvinyl formal from vinyl acetate monomers is as follows:

Vinyl acetate   Polyvinyl acetate   Polyvinyl alcohol   Polyvinyl formal

1.2.2.2 Blending Modification

Blending modification is the process of preparing a composite material that is macroscopically uniform by mixing two or more kinds of polymers. It usually includes physical blending, chemical blending, and physical-chemical blending. Physical blending is the blending in common sense, in which the chemical structure of large molecular chains in polymer blends is not changed; only the system composition and microstructure change. Chemical blending is the research category of chemical modification, such as the chemical reaction of polymers and interpenetrating polymer network. Physical-chemical blending is the blending process in which some chemical reactions occur, such as chain transfer reaction and strand exchange reaction, but as long as the reaction proportion is small, it generally also belongs to the research scope of blending modification.

With blending modification, polymer composites with different properties can be combined, which not only can improve the performance of polymers but also can make use of complementary properties of different polymer materials to prepare a new polymer material with excellent performance. In addition, blending expensive and relatively inexpensive polymer materials can reduce the production cost while not reducing or only slightly reducing performance. Blending modification therefore has obvious advantages such as simplicity and convenience, strong operability, and wide application range, and is one of the most widely used modification methods. For example, polypropylene (PP), one of the most important common plastics, has excellent comprehensive performance, and is one of the most promising thermoplastic plastics, but its poor toughness, especially brittleness under low temperature, limits its practical application. However, rubber and elastomer have high elasticity and good low-temperature performance. Blending rubber with PP can improve its impact performance and low-temperature brittleness. In a study on PP/EPDM (ethylene propylene diene monomer) composites, Zhang[12] found that the tensile strength and hardness of the composites decreased with an increase of EPDM content, while the impact strength increased. When the content of EPDM was 25%, the impact strength of the composite increased to six times that of pure PP.

1.2.2.3 Filling Modification and Fiber-Reinforced Composite Modification

Filling modification of polymer is the addition of solid additives with different composition and structure to the polymer matrix material to reduce costs or obviously change the performance of polymer products, which will improve the desired performance at the expense of other kinds of performance at the same time. Such an additive is known as a filler. Because these fillers are mostly inorganic powder, filling modification relates to performance difference and complementation of organic polymer and inorganic matter. This provides diverse areas of research and broad fields of application for filling modification.

In the filling modification system, carbon black for rubber reinforcement is the most typical example. It is this reinforcement system that promotes the development of the rubber industry. In the field of plastic, filling modification not only can improve performances, but it also plays an important role in reducing cost. Hong[13] filled superfine heavy calcium carbonate into polypropylene and obtained a composite material. The results showed that the tensile modulus and flexural modulus of the composites increased obviously with the increase of heavy calcium carbonate content. Polymer materials filled and modified with inorganic whiskers also belong to this scope, and certain advances have been achieved. Jiazhen Lv[14] cured potassium titanate whiskers and nylon-66 composite using epoxy resin with an elastic interlayer. The results showed that the impact strength of the composite increased by 132% compared with nylon-66 and pure nylon and the bending strength and tensile strength increased by 55% and 48%, respectively.

Fiber-reinforced composite modification often refers to using fiber-reinforcing filler or reinforcement material with a large length-to-diameter ratio such as glass fiber, carbon fiber, and metal fiber to enhance the mechanical properties of polymer materials. For example, if untreated carbon fiber is added to PVC, the highest tensile strength reached is 51 MPa when adding 12 phr. When the carbon fiber is treated with a coupling agent and then added to the PVC, the tensile strength increases to 90 MPa when adding 11.7 phr.[15]

1.2.2.4 Surface Modification

With the expansion of the development and application field of the polymer material industry, people began to pay attention to polymer surface properties, such as adsorption, adhesion, printing, antifogging, dyeing, electroplating, and lubrication. To meet these needs, surface modification of materials is required. Commonly used high polymer material surface modification methods include chemical and physical modifications. Physical modification includes surface mechanical modification, surface coating modification, surface vacuum plating, sputtering, injection, physical vapor deposition method, and so on. Chemical modification includes surface flame modification, solution treatment, discharge, ray radiation, electroplating, ion plating, graft polymerization, and chemical vapor deposition. Matured and widely used surface modification techniques include surface corona discharge treatment, flame treatment or heat treatment, solution treatment, radiation treatment, plasma treatment, surface plating, ion plating, and graft polymerization. At present, surface modification has become an important part of the surface of polymer materials.

1.2.2.5 Coextrusion Compound Modification

Coextrusion compound modification is the process in which a number of extruders are used to produce different molten material flows individually, which are then gathered in a complex head sink to obtain multilayer composite products. The procedure can combine multilayer materials with different characteristics in the extrusion process, so the product will have the characteristics of several different materials. It therefore makes it possible to achieve special requirements in performance and appearance, such as anti-oxygen and anti-moisture insulation, dye-ability, thermal insulation, thermal forming and heat adhesive ability, strength, rigidity, toughness, and other mechanical properties.

In addition, coextrusion compound modification can greatly reduce product cost, simplify processes, and reduce equipment investment. The composite process does not use solvents and therefore will not produce waste gas, waste water, and industrial residue. For these reasons, coextrusion composite modification is widely used in the production of compound fibers, films, plates, pipes, profiles, wires, and cables.

1.2.2.6 Recycling of Polymer Materials

Compared with traditional materials, polymers have the advantageous characteristics of light weight, easy moldability, beauty, and practicality. Because of these advantages, the polymer products are increasingly consumed in recent years and are widely used in all aspects closely related to people's lives.

The three main products considered representative of polymer material applications are plastic, rubber, and chemical fiber, and production of all of these is rising. In China, plastic products are concentrated mainly in agricultural plastic products, packaging plastic products, building plastic products, industry, transportation, and engineering plastic products. In the "eleventh-five-year-plan" period (from 2006 to 2010), the annual growth rate of the total output of plastic articles was 10%, the annual growth rate during 2010–2015 is estimated to be 8% in 2015, and plastic products output will reach 50 million tonnes. During the "eleventh-five-year-plan" period, the quantity demand of synthetic rubber in China will grow at an annual rate of 6%, and the annual production will reach 3.21 million to approximately 3.55 million tonnes in 2010. In

2006, the fiber yield of the world increased by 5.1%, reaching 74.70 million tonnes; in China, synthetic fiber (including filament fiber and staple fiber) production was 18,603,200 tonnes, with an increase of 11.82%. From these data we can see that millions of tonnes polymer products are in circulation in the market each year, bringing great convenience to people's lives, and raising living standards. But at the same time, along with the degradation and upgrading of products, the amount of waste converted by these products is self-evident. Therefore, the treatment of waste polymer materials has become a global problem.

Recycling of polymers is an important part of dealing with these issues and has received extensive attention. Numerous successes have been experienced, although at present, many problems and difficulties exist in polymer recycling. Physical recycling cost is very high as a result of the lack of market competitiveness. Some polymer wastes contain a large number of impurities and cannot be easily removed, or a variety of mixed materials cannot be easily separated. Generally, after melt processing the thermoplastic polymer will undergo thermal degradation to different degrees, which will decrease polymer molecular weight and broaden molecular weight dispersion to a value somewhat different from the raw material index. Thermosetting polymer is already a crosslinked polymer and cannot be dissolved or melted, which will increase the difficulty of recycling.

With social and economic progress, the application of polymer materials will grow and the corresponding waste will increase rapidly. Therefore, making full use of resources and reducing environmental pollution are the final objectives of polymer material recycling.

1.2.3 Development Trends in Modification of Polymer Materials

With the growing expansion of polymer material application fields, the variety of polymers has increased rapidly.

The structure of polymer industrial raw materials has not changed much in the 21st century, but there have been greater improvements in quality and performance. Sustainable functional polymers and high-performance composite materials are being developed at an annual rate of more than 10%. Polymer alloy, polymer composite material, liquid crystal polymer material, polymer nanometer material, and other new polymer materials have been successfully developed.

Modifications of high polymer materials are briefly summarized according to the following aspects.[3,4]

1. The developing trends of general synthetic polymeric materials are high performance, multifunctionality, low cost, and environmental sustainability. For example, plastics modification is the addition of relevant modifying agents to mass general resins to improve or add functions to meet application requirements under special conditions in the aspects of electricity, magnetism, light, heat, anti-degradation, flame retardation, and mechanical properties through physical, chemical, or mechanical methods. Plastics filling, blending, and reinforcing modification techniques have gone fairly deeply into the raw materials and forming processes of all plastic products; impacted the traditional plastic industry; and greatly promoted the development of the machinery manufacturing industry, auxiliary industry, and nonmetallic minerals industry. Modified polypropylene (PP) has become one of the most popular products. Glass fiber- and mineral-reinforced PP has been used in automotive parts and has impacted the market of polycarbonate (PC) blends. At present, 90% of glass fiber-reinforced PP in the world is used for cars. In addition, modified PP is entering the field of traditional polyamide (PA), such as pumps and fan blades. Modified plastics therefore is a plastic industry field

with high-technology products, is involved in a wide range of applications, and can create significant economic benefits.

2. Functional polymer materials, special polymer materials, and engineering plastics are developing rapidly.

Polymer materials with optical, electric, and magnetic properties will play an important role in the development of the whole information technology field in the 21st century. Therefore, the research and development of organic polymer photoelectric information functional materials, such as organic polymer display materials, in particular electroluminescent materials, ultra-high-density polymer storage materials, and polymer biosensing materials, is getting more attention. In addition, the design, simulation, calculation, synthesis, and assembly of new functional polymer materials; the construction of molecular nanostructures; the assembly and self-assembly of polymers; the application research in molecular electronic devices; and so on are also gaining increased consideration.

For example, the JSP Company in Japan developed a non-crosslinking foaming polyethylene (PE) plate with a permanently antistatic polymer by using a unique microdispersion technology employing a macromolecular antistatic agent. The plate allows surface electric resistance to remain stable in the range of 1011–1012 Ω and has the advantages of not destroying the original impact resistance of materials, saving energy and promoting recyclability. This product is used in computer-related products with antistatic performance and food cushioning packaging materials to avoid surface adsorption of dust caused by static electricity and weak current caused by pollutants. In addition, the national engineering research center of engineering plastics in China produced antibacterial materials and products through unique antibacterial masterbatch technology and launched the first domestic antibacterial household products by merging with the Haier group.[16]

3. Coordinated development with energy and environment-related organic polymer materials is attracting more and more attention.

   Synthetic polymer materials based on petroleum sources have been applied in a large scale and have brought us convenience, but at the same time they also brought environmental pollution, and we will face the threat of gradual depletion of petroleum resources after 50 years. Therefore, in the future the main chemical raw materials for natural polymers will likely be based on renewable animal, plant, and microbe resources. The most abundant resources are cellulose, lignin, chitin, starch, various animal and plant proteins, and poly-saccharides. They have multiple functional groups and can become new materials through chemical and physical modifications, and also become chemical raw materials by degradation to monomers or oligomers through chemical, physical, and biological methods.

4. Research into the processing field of polymer materials continues to expand and deepen.

   The final usability of polymer materials depends largely on the material's morphology after formation. The morphology of polymer includes crystallinity, orientation, and so on, and multiphase polymers also include phase morphology (such as sphere, sheet, rod, fiber, etc.). The morphology of the polymer products is determined mainly by the effect of the complex temperature field and the external force field during processing. Therefore, studies on morphological formation, evolution, regulation, and eventually "structuring" of polymer materials under the action of an external force field during processing and development of new methods for polymer processing and molding have important significance for the basic theory research of polymer materials and development of clean compound polymer materials with high performance, multifunctionality, and low cost.

5. The intersection of polymer materials science with other disciplines continues to strengthen.

From its earliest days, polymer science has been a multidisciplinary field. The intersection of polymer material science with biological science, biological engineering, chemistry, physics, information science, and environmental science has promoted the development of polymer material science itself and expanded its scope of application. For example, most biological tissues exist in nature in the form of soft material (polymer hydrogel). Polymer hydrogel is the system most similar morphologically to biological and plant structure among compounds known to date. In particular, polymer hydrogel can make a soft reaction to external environmental changes and show more typical soft matter properties and bionic intelligent characteristics. Researchers at Wuhan University[17] studied the structure of natural polysaccharides and cellulose physical gel and the gelation process using a variety of advanced technologies combining polymer, biology, physics, and other disciplines.

Biomimetic polymer material that can emulate the structure of an organism or a particular feature is an important way to develop biological materials.

In a word, modification of polymer materials can give polymer materials new properties at lower cost, greatly improve the performance of materials, or provide new functions, further broadening the application field of polymer materials and greatly increasing the industrial application value. In the future, the modification of polymer materials will still be one of the research hotspots in the polymer material science and engineering fields.

References

1. Rumin Wang, Shuirong Zheng, Yaping Zheng. *The process of polymer matrix composites*. Beijing: Science Press, 2004.

2. Xiaomin Cheng, Chuli Shi. *The introduction of polymer materials.* Hefei: Anhui University Press, 2006.
3. Yongmei Chen, Chun Li. *Polymer basis.* Beijing: Science Press, 2009.
4. Shen Wang. *Modification technology of polymers.* Beijing: Chinese Textile Press, 2007.
5. Jing Guo. *Modification of polymeric materials.* Beijing: Chinese Textile Press, 2009.
6. Yihua Cui, Xinxin Wang, Zhiqi Li et al. Fabrication and properties of nano-ZnO/glass-fiber-reinforced polypropylene composites. *Journal of Vinyl and Additive Technology,* 16(3):189–194, 2010.
7. Qiuju Sun, Baoling Zang, Baoyan Zhang et al. Synthesis and characterization of side-chain liquid crystalline polysiloxanes containing sulfonic acid ionic groups. *Journal of Northeastern University,* 26(4):275–279, 2005.
8. Baoyan Zhang, Qiu Sun ju, Mei Tian et al. Synthesis and mesomorphic properties of side-chain liquid crystalline ionomers containing sulfonic acid groups. *Journal of Applied Polymer Science,* 104(5):304–309, 2007.
9. Qiuju Sun, Baoyan Zhang, Danshu Yao t al. Miscibility enhancement of PP/PBT blends with a side-chain liquid crystalline ionomer. *Journal of Applied Polymer Science,* 112(5):3007–3015, 2009.
10. Wei Tian, Yunfei Gu. Application and development of electroless silver plating. *Electroplating & Pollution Control,* 30(3):4–7, 2010.
11. Ruihai Li, Yi Gu. Study on the electroless nickel plating on polystyrene plastics. *Journal of Sichuan University (Engineering Science Edition),* 36(6):63–66, 2004.
12. Zhihong Zhang. Study on the properties of polystyrene/elastomer composites. *Plastics Manufacture,* 32(9):100–102, 2007.
13. Lianzhou Hong, Shu Bi, Ping Ning. Ultrafine ground calcium carbonate filled polypropylene. *Plastics,* 25(5):31–35, 2006.
14. Jiazhen Lu, Xiaohua Lu. Elastic interlayer toughening of potassium titanate whiskers–nylon66 composites and their fractal research. *Journal of Applied Polymer Science,* 82(2):368–374, 2001.
15. Heng Chu, Chuner Wang, Chunqing Li. Study of tensile strength of PVC reinforced by carbon fiber. *China Plastics Industry,* 36(s):100–103, 2008.
16. Bozhang Qian. Development of reinforced world plastics. *World Plastics,* 28(7):36–39, 2010.
17. Lihui Weng. *Studies on characterization of polymer gels and their intelligent behavior.* Wuhan: Wuhan University, 2004.

Chapter 2

General Principles of Filling Modification of Polymer Materials

Filling modification of polymer materials is the addition of solid additives with different compositions and structures to the overall polymer, giving polymer products certain new performances or reducing the cost without decreasing or with minimum sacrifice of performance, significantly improving the desired performances, further broadening the application field of polymer materials, and greatly improving the polymer industrial application value. Such additives are called filling agents or fillers.

Adopting inorganic mineral fillers for filling modification of plastics began in the late 1970s in China. The earliest inorganic filler masterbatch was atactic polypropylene (APP), which uses APP as the carrier resin and 400-mesh heavy calcium carbonate (particle size 0.037 mm) as the filler.[1] At that time, this masterbatch was used mainly in the manufacture of pipes and sheets and played a positive role in saving resin and reducing cost. Subsequently, the carrier resin of the filling masterbatch was gradually replaced by polyethylene or other polymer materials, and the filler was also expanded from a single heavy calcium carbonate to light calcium carbonate, talc, mica, kaolin clay, silica, white carbon black, titanium dioxide, fly ash, red mud stone, diatomite, wollastonite, and glass beads.[2] Filling plastics with inorganic mineral has the following advantages.[3]

1. *Cost reduction.* Because the price of mineral powders is lower than that of the synthetic resin by 1/10–1/15, a suitable amount of filling can reduce the price of plastics.
2. *Reduction in the consumption of synthetic resin.* Synthetic resin is a petrochemical product, so reducing the consumption of petroleum products aligns with the requirement for sustainable development strategies in today's society with an increasing focus on energy aspects.[4]
3. *Improvement in the performances of the plastic products.* Mineral filling can improve the comprehensive performances of products, such as hardness, elastic modulus, dimensional stability, thermal stability, printing effect, and so on.[5]
4. *Improvement of environmentally-friendly performance.* Mineral filling can improve the photodegradation properties of plastic products, so it has important implications for reducing white pollution.[6]

2.1 Effect and Variety of Fillers

2.1.1 Effect of Fillers

Fillers are different from additives that are commonly used in polymer processing, such as paints, heat stabilizers, flame retardants, and lubricants, and also are different from other liquid fertilizers. Fillers used in polymer modification have the following characteristics.

1. Solid matter with certain geometric shapes could be either inorganic matter or organic matter.
2. Fillers usually do not react with the matrix polymer.
3. Fillers have a low mass fraction in the polymer filling system, generally 5%–30%.

In a polymer filling system, owing to differences in their chemical composition, geometric shape, distribution of particle sizes, surface morphology, dispersion of fillers in polymer, and interface structure, fillers play different roles in filling modifications of polymers, such as incremental packing effect, matrix enhancement, toughening, and lending special functions, and so on.[7,8]

2.1.1.1 Incremental Packing Effect

Because the price of packing material is generally lower than that of the base material, packing of the same price has a much larger mass or volume than the matrix, so it can give the filling system a larger mass or volume and therefore achieve the incremental packing goal and reduce cost.

The ratio of the actual volume of fillers to the total volume of the filled polymer is called the packing coefficient. The higher the packing coefficient, the larger is the amount of filler that can be filled. Spherical particles have the greatest symmetry and the smallest surface area and therefore have the highest packing coefficient. Packing with an irregular geometry causes an inability to form an orderly arrangement, so its effective space utilization is low, leading to small packing density and low maximum filling content.

As an example, calcium carbonate powder is an inexpensive filler with a high degree of whiteness that is widely applied to the processing of polyolefin products. The content of calcium carbonate powder is as high as 80% in the masterbatch. Filling with calcium carbonate dramatically reduces the cost of the product, but too much filling will lead to embrittlement of products and shorten the service life, especially because of pulverization in sunlight, which easily leads to material damage. Surface-modified heavy calcium carbonate was filled into polyethylene material by Dong et al.[3] The results show that processing performance gradually declines with an increase of the proportion of packing; when packing is higher than 30%, a film break can easily occur in the process of blowing film products.

On the basis of a simple stretching model, Nielsen[9] deduced the tensile yield equation when the filler and polymer have no adhesion whatsoever. The filled plastic tensile strength can be roughly calculated by Formula 2.1.

$$T_r = T_p (1 - V_f) + T_{fp} V_f \qquad (2.1)$$

where

T_r = the tensile strength of the filling compound, MPa
T_p = the tensile strength of the pure polymer, MPa
T_{fp} = the interface tensile strength between the filler and the polymer, MPa
V_f = the filling volume fraction of the filler, %

According to the theoretical calculation of the model, when the volume of the filler is greater than 75%, the packing particles can make direct contact with each other and the tensile strength decreases to almost zero.

Therefore, in the practice of polymer filling modification, any extra effects of valuable performances of polymer materials must be prevented, and an integrated balance should be achieved among the contradictory effects.

2.1.1.2 Enhancement Function

With technological developments and economic considerations, the application fields of polymer materials become wider, consumption grows, and the technical performance requirements become increasingly higher. General plastics, such as polyethylene (PE), polypropylene (PP), polyvinyl chloride (PVC), acrylonitrile butadiene styrene (ABS), with the advantages of low price and convenient molding, have broad application prospects. But because of the low structural strength, their characteristics, such as low temperature resistance, aging resistance, and fatigue resistance, make it difficult to meet the demand of engineering applications. Therefore, using resin as a matrix, adding fillers could improve performances such as mechanical strength, heat resistance, molding shrinkage, and linear expansion coefficient. These fillers could be termed reinforcing materials.

Reinforcing material usually has high strength and modulus; when compounded with a resin matrix, it can enhance the mechanical effect or other performance characteristics of the resin. Glass fiber, carbon fiber, asbestos fiber, boron fiber, aromatic polyamide fiber material, and so forth are commonly used as reinforcing materials. For example, long glass fiber was continuously filled in polyphenyl ether to perform enhancement modification. The results show that the strength of the composite material system is visibly increased after the addition of glass fiber. When the glass fiber content is 20%, the tensile strength and the bending strength of the composite material system are 90 MPa and 103 MPa, respectively, and increased by 61% and 91% respectively compared to the pure polyphenyl ether system.[10] As another example, the surface effect, small size effect, and quantum tunnel effect of nanoparticles give them high surface activity, large specific surface area, and a series of special physical and chemical properties. The tensile strength of the blend PP/PVC (40/60) increases gradually with a rising addition of nano-montmorillonite. When the content of nano-montmorillonite is 5 parts per hundreds of resin, the tensile strength of the blend reaches a maximum value and increases by 22% compared to the pure PP/PVC blend.[11]

2.1.1.3 Toughening Effect

In polymers used as structural materials, strength and toughness are two important mechanical performance indexes. Toughening changes the rupture of a polymer from a brittle fracture to ductile fracture and gives the polymer a higher break elongation under stretch, making it harder to damage when hit, or it can absorb more fracture energy once destroyed.

At present, the widely used toughening method uses an elastomer as the toughening agent and a mature toughening mechanism has been developed. Although the toughness of a polymer could be increased significantly by using an elastomer, the modulus, strength, and thermal deformation temperature of the blend noticeably decrease, and the cost is relatively higher. However, the impact resistance of material could be increased by adding inorganic rigid particles, and at the same time processing fluidity and heat deformation could also be improved. Furthermore, using inorganic rigid particles does not reduce the tensile strength and rigidity, and can reduce costs and expand the scope of the polymer's application.[12,13] For example, surface-modified calcium carbonate was filled into the polypropylene material by Zhu et al.[14] The result shows that Young's modulus and notched impact strength of the composite material increase at the same time and reach the most significant toughening effect when calcium carbonate has a mass fraction of about 40%.

2.1.1.4 Imparting Special Functions to the Material

With the development of filling modification technology and a deepening understanding of fillers, reasons for using them, especially the functional fillers, have expanded from pure cost reduction to a provision of some performance characteristics to products, such as electrical property, magnetism, electric wave absorbency, ultraviolet resistance, antibiosis, and other special functions. Seeking new functional and high added-value fillers has become mainstream. For example, carbon nanotubes (CNTs) are tubular fibers curled up by one or several graphite layers, whose interiors are empty and only a few to dozens of nanometers in diameter. As conductive additives, carbon nanotubes have a unique advantage. Conductive plastics produced by carbon nanotubes have achieved commercialization and gained more and more applications in areas such as the electronic industry, automobile industry, and so on.[15] Another example is potassium titanate whiskers, which are an inorganic monocrystal fibrous material with a certain aspect ratio. They have properties such as chemical stability, excellent thermal insulation and wear resistance, and so on. When surface-modified potassium titanate whiskers are filled into polyether ether ketone (PEEK), the antifriction ability and wear resistance of the material increase significantly. The wear resistance of the material increases by 2.64 times under a 300 N force.[16]

2.1.2 Classification of Fillers

The properties of fillers such as geometric shape, particle size and distribution, physical and chemical properties, and so on will directly affect the material performances of the filling modification system. Fillers are classified mainly into the following kinds.

2.1.2.1 According to the Composition of the Filler

Fillers can be divided into the following types according to their compositions.

1. Oxide categories: white carbon black, diatomite, alumina, and so forth
2. Hydroxyl categories: calcium hydroxide, magnesium hydroxide and aluminum hydroxide, and so forth
3. Carbonate categories: calcium carbonate, magnesium carbonate, and so forth

4. Sulfate categories: calcium sulfate and barium sulfate, and so forth
5. Silicate categories: talcum powder, bentonite, mica powder
6. Nitrides categories: silicon nitride and aluminum nitride
7. Carbon categories: carbon black, carbon nanotubes, carbon fiber
8. Other categories

2.1.2.2 According to the Shape of the Filler

According to their shapes, fillers can be divided into fibrous, acicular, flake, and granular fillers. Glass fiber, carbon fiber, aramid fiber, and so forth belong to long fiber materials. The whisker fillers that are a focus of this book can be considered acicular materials. Whiskers have a powdery appearance but are acicular or fibrous under a microscope, with lengths of a few microns to hundreds of microns. Figure 2.1a and b shows scanning electron microscopic (SEM) images of calcite type (particle) and needle-like calcium carbonate, respectively.

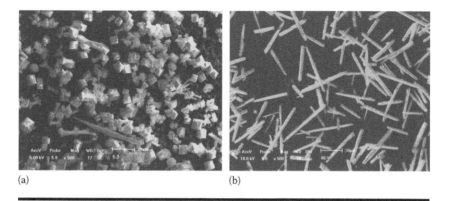

(a) (b)

Figure 2.1 **SEM images of calcium carbonate. (a) Calcite type. (b) Acicular type.**

2.1.2.3 According to Filling Goals and Applications

Fillers can be divided into the following three kinds according to the filling goals and applications:

1. Incremental packing, such as calcium carbonate, talcum powder, and white carbon black.
2. Enhanced and toughening packing, including glass fiber, carbon fiber, aramid fiber, all kinds of whiskers, and so on.
3. Special functional packing, including filling agents with electric conduction, thermal conductivity, piezoelectricity, magnetism, shock absorption, flame retardance, wave absorption, and so forth. For example, carbon fiber and carbon black fillers have electrical conductivity; alumina and beryllium oxide fillers have thermal conductivity; filling various kinds of ferrites can increase the magnetic and wave absorbing properties; and so on.

2.2 Effects of Filler Properties on the Filling System

The filler type, particle size, shape, performance, and compatibility with polymer material all have some effects on the performance of the filled polymer materials.[7,8]

2.2.1 Varieties of Packing

Different types of fillers have different effects on the filling system. The choice of packing is often based on the end use of the product. For example, acid-proof products generally do not contain calcium carbonate; to increase flame resistance, aluminum hydroxide or another substance is generally added as a filler. The nature of the filler is closely related to its chemical composition. That is to say, the packing provides some functional role; in most cases it is the chemical composition of the filler at work. However, chemical compositions are not the only influencing factors; physical properties also have an effect.

2.2.2 Size of Packing

In general, the shapes of filling particles are very irregular, so it is difficult to study the effect of geometry on the properties of the filling system directly. The aspect ratio (length-to-diameter ratio) therefore is usually measured. The influence of the aspect ratio of filler particles on filled performance often is attributed to the influence of the surface area or specific surface area of filler particles on the filling system. There are also some qualitative conclusions. For example, adding fillers with a small aspect ratio into flexible thermoplastic plastics can increase the compression strength, while adding fillers with a large aspect ratio can possibly increase tensile strength slightly.[17] In addition, with the same filling mass fraction, the smaller the particle size of the filler, the higher is the viscosity of the filling system. Also, the smaller the particle size, the easier it is for particles to combine with each other. Fillers in an aggregation state hinder the liquidity of the filling system.

2.2.3 Surface Property of Packing

A transitional zone with a certain thickness is formed between two phases when any two phases come in contact, and this transitional zone is usually called an interface. If one phase is gas, this interface is usually referred to as the surface. The surface area of unit mass packing is known as the specific surface area. According to the thermodynamic theory, in the reversible condition of constant pressure, temperature, and composite, with an increase in unit surface area, the added value of Gibbs free energy is known as the specific surface Gibbs free energy. It is also referred to as the surface free energy, expressed in J/m^2.

The surface property of a filler depends on its chemical composition, crystal structure, adsorption material, presence of surface pores, and so forth. The smaller the particle size, the larger is the specific surface area. For particles with the same volume, the surface area is associated not only with the geometric shape of the particles, but also with the surface roughness. The rougher the surface, the larger is the surface area. In addition, the surface free energy of the filler is related to the difficulty of dispersion of the filler in the polymer matrix. At a certain specific area, the larger the surface free energy, the easier it is for the filler to aggregate and the more difficult to disperse. As a result, reducing surface free energy is one of the main goals when modifying a filler surface.

2.2.4 Piling Form of Packing

In a composite system, packing is unlikely to exist as single particles, especially when the surface area is large; it always exists in a certain state of aggregation. Different size distributions of filler particles lead to different pile forms of packing in a composite system: the largest dense packing system and the least dense packing system. The former is due to the scatter particle size distribution and the latter is the result of relatively concentrated particle sizes.

The accumulation measurement method designed by C. C. Furnas is based on the determination of void volume of the particles with uniform shape. The total volume of the packing system depends mainly on the packing of coarse particles, because fine particles can be placed in the gaps among coarse particles; therefore the total volume is not changed at this time. Finer particles are then added and may be placed in the gaps among fine particles or the gaps among fine particles and coarse particles. The gaps brought by finer particles are filled by even finer particles. Under the condition of unaffected total volume, this kind of piling with so many grains forming part of the stack is called the largest dense packing.

Conversely, the least dense packing can be seen if the filler is stacked by particles with the same size or shape. It contains more space and therefore the whole stack system occupies a larger space. For example, it is impossible for spheres with the same diameter to be placed in the gaps among spheres with the same diameter; and filler particles with the same needle shape are often mixed and piled in a disorderly manner into a loose group because of different particle orientations in a static state.

2.2.5 Physical Properties of the Packing

Physical properties of the packing itself, such as density, hardness, heat resistance, electrical property, magnetic and optical properties, water absorption, and oil absorption performance are the basic requirements in choice of packing. A filler with high hardness can improve wear resistance of the filling system, but meanwhile the wear of the plastics processing equipment is also larger. The heat transfer coefficient of graphite is significantly higher than that of polymers, and this lays a foundation for preparing graphite filler material with corrosion resistance and high conductivity.

2.2.6 Thermochemical Properties

A polymer material burns easily, and decorative materials that are made of polymer materials, such as PE, PP, and some natural polymer such as cellulose and so forth frequently pose fire hazards. In addition, a great deal of smoke is produced when some materials such as polypropylene and PVC are burned. Because most packing materials are incombustible, adding packing could reduce the concentration of combustible materials and delay matrix combustion. Shang added the fibrous $Mg(OH)_2$ and globular $Al(OH)_3$ to polypropylene to prepare a composite material.[18] Their result shows that the oxygen index of the composite material with fibrous $Mg(OH)_2$

and globular $Al(OH)_3$ fillers is 27.3, which meets the requirement of flame retardance. At the same time, under the same distribution ratio of each component of $Mg(OH)_2$–$Al(OH)_3$ compound flame retarder, the tensile strength and impact strength of the sample with fibrous $Mg(OH)_2$ filler are larger than that with globular $Mg(OH)_2$ filler. To a certain extent, fibrous $Mg(OH)_2$ has a reinforcing effect on glass short fibers.

2.3 Surface Modification of Inorganic Fillers

Because of their unique physical and chemical properties, inorganic fillers can improve the mechanical properties, processability, thermal performance, and so on, of polymers. Therefore, the application of inorganic fillers in polymers has been developing rapidly. However, because of their good hydrophilicity, inorganic fillers have poor compatibility with hydrophobic organic polymer materials; consequently, inorganic fillers will be unevenly dispersed in the matrix if added directly. Furthermore, the large particles will become stress concentration points in the polymer material and hence become weak links in materials. This not only limits the amount of filler in the polymer, but also seriously affects the mechanical properties of the composite. Therefore, the surface of inorganic fillers must be modified.

Surface modification is the process in which the surface of inorganic particles reacts chemically or physically with a surface modification agent, thereby improving the chemical and physical properties of the packing surface. The dispersity, stability, compatibility with polymer, and processing fluidity of surface modified inorganic fillers have been improved markedly, in accord with the requirements of application fields. Xu et al.[19] compared the effects of the unsaturated polyester resin filled by nanometer TiO_2 with or without surface treatment. The results show that the filling effect of surface-modified nanometer TiO_2 is obviously better than that without surface treatment. The reinforcing and toughening effects of the composite are best when surface-modified TiO_2 has a mass fraction of 4%. The tensile strength, bending strength, elongation at break, and impact strength of the composite are increased by 102%, 184%, 125%, and 91%, respectively.

2.3.1 Surface Modification Methods of Packing

According to the implementation of modification methods, surface modification methods of inorganic packing are of the following kinds: local chemical modification, mechanochemistry modification, surface coating modification, encapsulation modification, high-energy surface modification, and so forth. Among these methods, local chemical modification and surface coating modification are widely used.

2.3.1.1 Local Chemical Modification

Local chemical modification is a method that uses chemical reactions of the functional groups on the surface of particles and the surface modification agent to achieve the modification. Before local chemical modification, a modification agent should be chosen according to the surface functional groups of inorganic particles. Main local chemical modifications include the following kinds: coupling agent modification, surface grafting modification, modifications through reactions with fatty acid or alcohol, and so forth.

2.3.1.1.1 Coupling Agent Modification

A coupling agent is a substance with an amphoteric structure. At one end of the coupling agent are polar groups that can react with inorganic particles and form strong chemical bonds; at the other end are nonpolar groups that can react chemically or interact physically with organic matter, thus building molecular bridges between inorganic fillers and organic polymers and enhancing the interaction between the polymer matrix and the packing, thus improving the performance of the products.

There are many types of coupling agents. The coupling effects of various coupling agents are closely related to the composition of the material that will be modified and the structure of the resin. According to the molecular structure of the packing, the appropriate choice of coupling agent can improve the comprehensive performance of composite materials.

1. *Silane coupling agent.* The silane coupling agent was the first coupling agent discovered and used most widely. The basic structure of the silane coupling agent is

$$R_n - SiX_{(4-n)}$$

where

R stands for organic hydrophobic groups, which have strong affinity and reaction capacity with polymer molecules, such as methyl, vinyl, amide, naphthenic base, sulfur, propyl acryloyl oxygen, and so forth.

X stands for the groups that can hydrolyze when they come in contact with a water solution, moisture in the air, or moisture adsorbed on the surface of inorganic materials and have good reactivity with the inorganic surface.

Typical X groups include alkoxy, aryloxy, acyl, chlorine, and so forth; methoxyl and ethoxy are commonly used.

Chlorosilane produces the corrosive by-product hydrogen chloride in the coupling reaction process and thus should be used judiciously.

In China, there are many types of silane coupling agents, such as KH550, KH560, KH570, KH792, DL602, DL171, and so forth. The names and structures of common silane coupling agents are shown in Table 2.1.

Table 2.1 Names and Structures of Common Silane Coupling Agents

Trade Name	Chemical Name	Structural Formula
KH550	(3-Aminopropyl)triethoxysilane (APTES)	$H_2N(CH_2)_3Si(OCH_2CH_3)_3$
KH560	γ-Glycidyl ether trimethoxypropyl silane	$CH_2\!-\!CH\!-\!CH_2O(CH_2)_3Si(OCH_3)_3$ $\diagdown O \diagup$
KH570	3-Methacryloxypropyltrimethoxy silane (MAPTMS)	$CH_2\!=\!C\!-\!CO(CH_3)_3Si(CH_3)_3$ $\underset{H_3C}{\mid}\ \underset{O}{\parallel}$
KH580	γ-Thiopropyl triethoxy silane	$HS(CH_2)_3Si(OCH_2CH_3)_3$
KH590	(3-Mercaptopropyl) trimethoxysilane	$HS(CH_2)_3Si(OCH_3)_3$

Silane coupling agents are often prepared in a solution with a certain concentration before use, and then the packings are dealt with. The silane coupling agent could be diluted with a very dilute solution (mass fraction 0.0005–0.02) composed of water and ethanol. It can also be diluted only with water, but to improve the solubility and promote hydrolysis, the silane coupling agent should be added to an acetic acid aqueous solution with a mass fraction of 0.001. The silane coupling agent could also be diluted with a nonaqueous solution such as methanol, ethanol, propanol, or benzene solution, or it can be used directly. It is recommended that the silane coupling agent dosage is 0.8%–1.5% of the material being processed, commonly 0.5%–3%. The effect is not proportional to the dosage of silane coupling agent. A surplus of a coupling agent may cause performance degradation, while the coating is not complete if the concentration is too small. Hence the best concentration has to be found experimentally to achieve both economy and effectiveness.

There are several major mechanisms of action of silane coupling agents, such as chemical combination theory, physical adsorption theory, hydrogen bond formation theory, reversible equilibrium theory, and so forth. The theoretical model proposed by Arkies is considered to be one of the most realistic theories. The reaction process of silane coupling agents with maleic anhydride grafted polypropylene (MPP) and the surface of inorganic substances according to this mechanism is shown in Figure 2.2. The silane coupling agent first hydrolyzes with air moisture and then dehydrates to form oligomers. Hydrogen bonds are formed between the oligomer and the hydroxyl on the surface of inorganic substances, and some covalent bonds are formed by a dehydration reaction after heat drying; therefore the surface of the inorganic substance is covered by a silane coupling agent.[20]

According to the above mechanism, the surface of inorganic substances cannot play the corresponding role or produce the expected effect if there is no hydroxyl group.

In addition to the traditional silane coupling agent, there are some novel silane coupling agents, such as organic silicone peroxide coupling agents. The peroxyl radicals of an organic silicon peroxide coupling agent are

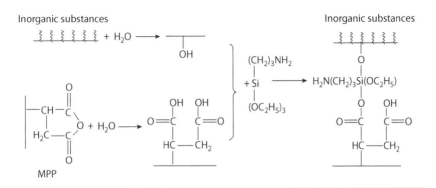

Figure 2.2   **A schematic diagram of the reaction of the silane coupling agent (KH550) with maleic anhydride grafted polypropylene and an inorganic substance on the surface.**

easily decomposed into free radicals with high responsiveness after being heated. Therefore, they not only can be used as coupling agents between organic and inorganic matter, but also can make two of the same or different organic couplings, and can be used as coupling agents of nonpolar organic matter (such as polyolefin, silicone rubber, etc.). Common organic silane coupling agents cannot accomplish these.[21]

2. *Titanate coupling agents.* In the 1970s, the titanate coupling agent was successfully developed in the United States by the Kenrich Oil Chemical Company. To date, a variety of titanate coupling agents have been developed and are used mainly in polyolefin plastics, coatings, adhesives, and so on that use carbonate, sulfate, and metal oxide as fillers. The structure of a titanate coupling agent is as follows:

$$\underset{\text{Hydrophilic phase}}{\underleftarrow{\hspace{3cm}}} \quad \underset{\text{Hydrophobic phase}}{\underrightarrow{\hspace{3cm}}}$$

$$\left(R_1 - O\right)_{\overline{m}} \longrightarrow \mathrm{Ti} - \left(O - X - R_2 - Y\right)_n$$

where

$1 \leq m \leq 4,\ m + n \leq 6$

R_1 = short carbon chain alkanes

R_2 = long carbon chain alkanes (aromatic hydrocarbons)

X = groups containing C/N/P/S, and so on

Y = hydroxyl, amino groups, epoxy group, and double bond, and so on

In the late 1970s, research on these compounds was also conducted in China, and a series of products of titanate coupling agents named Nan Da Zao (NDZ) were successfully developed. At present, there are more than 70 types of titanate coupling agents. According to the different coupling mechanism and molecular structure of coupling agents, the titanate coupling agents are often divided into four categories: chelating type, coordination type, pyro-phosphate type, and single alkoxy type. Common varieties of titanate coupling agents are shown in Table 2.2.

The molecules of titanate coupling agents have hydro-lyzable groups, especially the single alkoxy titanate cou-pling agent, and generally the fillers are not required to contain water. Pyrophosphate molecules have a pyrophos-phate base that can react with water, so they are suitable for packing containing crystallized water and no free water. Chelating molecules are relatively stable because they contain five-membered rings and do not react with free water, only with hydroxyl groups on the surface of the packing; they are therefore suitable for packing con-taining free water.

The choice of titanate coupling agent has a great deal to do with the inorganic filler. In general, the packings

Table 2.2 Common Varieties of Titanate Coupling Agents

Varieties	Trade Name	Chemical Name
Single alkoxy	NDZ-101	Isopropyl dioleic(dioctylphosphate) titanate
	NDZ-105	Isopropyl trioleyl titanate
	NDZ-130	Titanium triisostearoylisopropoxide
	NDZ-102	Isopropyl tri(dioctylphosphate)titanate
	NDZ-109	Titanium tris(dodecylbenzenesulfonate) isopropoxide
Pyrophosphate	NDZ-201	Isopropyl tri(dioctylpyrophosphate) titanate
Chelating	NDZ-311	*Bis(P,P-bis-*ethylhexyl diphosphato) ethanediolato titanate
Coordination	NDZ-401	Tetraisopropyl di(dioctylphosphate) titanate

that are more suitable for titanate coupling agents include calcium carbonate, barium sulfate, titanium dioxide, mica, talcum powder, kaolin, and so forth, whereas packings such as aluminum oxide, glass fiber, white carbon black, and so forth are not quite suitable for titanate coupling agents. The concentration of titanate coupling agent is usually between 0.1% and 2.0%. The thinner the inorganic filler, the larger is the concentration, and the optimal concentration of the titanate coupling agent varies for different packings.

Thermoplastic polymers filled by inorganic fillers that are modified by a titanate coupling agent have low forming temperature and short forming time, and their impact strength and mechanical properties can also be improved. In addition, the dispersibility of paints and coatings activated by the titanate coupling agent is improved, and therefore these paints and coatings can impart a bright color and smooth surface to the products. For example, adding carbon black, SiO_2, $CaCO_3$, metal oxide activated by a titanate coupling agent to the thermoplastic plastic and rubber can reduce the system's viscosity and improve toughness and mechanical properties. Under the premise of keeping elongation and improving the low temperature toughness of the PVC, a titanate coupling agent can decrease the consumption of PVC resin or increase the concentration of inorganic filler in the system.[22]

3. *The aluminate coupling agent.* Similar to a titanate coupling agent, an aluminum coupling agent can obviously drop the adhesion of inorganic fillers. At the same time, an aluminum coupling agent can greatly improve the amount of packing in an organic substrate. There are two types of active groups in an aluminum coupling agent: One group can react with the surface of inorganic filler, and another can interact with resin, and thus coupling

action is produced between inorganic fillers and the matrix resin. The general chemical formula is

$$(C_3H_7O)_x Al(OCOR)_m (OCORCOOR)_n (PAB)_y$$

where

$$x + m + n = 3; \; y = 0\text{--}2$$

Aluminum coupling agents are relatively inexpensive, with a cost that is only half that of titanate coupling agents, and have features such as light color, absence of toxicity, small flavor, and ease of use, and so forth and their thermal stability is also better than that of titanate coupling agents. In addition, with high reactivity with the surface of inorganic fillers, aluminum coupling agents are comparable with titanate coupling agents in improving physical properties, such as impact strength, thermal deformation temperature, and so forth of the plastics modified by an inorganic filler. The calcium carbonates activated by the aluminum coupling agent have the following characteristics: low moisture absorption, low oil absorption, smaller mean particle size, easy dispersion in organic medium, high activity, and so forth. They are widely used in filling modification of plastics, such as PVC, PE, PP, Polystyrene (PS), and so on. They not only can ensure the processing performance and physical performance of products, but can also improve the content of calcium carbonate to reduce the product cost.

4. *The double metal coupling agent.* A double metal coupling agent has the ability to introduce organic functional groups to two inorganic scaffolds, so it has performances that no other coupling agents have. For example, the processing temperature of double metal coupling agents is low, so they can interact with packing at room

temperature; double metal coupling has the advantages of elevating the coupling reaction rate and good dispersibility, so these agents can make it possible for modified inorganic fillers to mix easily with polymer; they also can improve the content of inorganic fillers to reduce product cost.[23,24]

An aluminum–zirconium organic metal complex coupling agent with the brand name of Cavedon Mod is one new type of coupling agent developed by American Cavedon chemical companies in the early 1980s. It was synthesized from zirconium oxychloride, chlorohydrin aluminum hydrate, propylene glycol, carboxylic acid, and so forth. Its molecular structure is shown below.

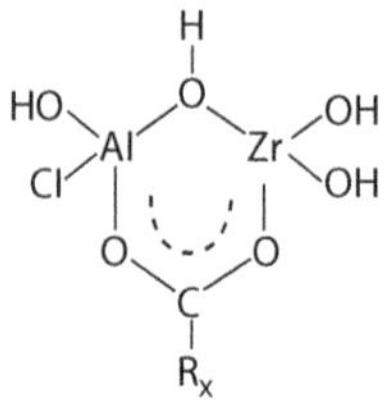

Among them, according to the needs of different types of resin, the R_x coordination group can be amino, carboxyl, hydroxyl, sulfur, methyl acrylic acid groups, and so on. The molecular structure of the aluminum–zirconium organic metal complex coupling agent contains two inorganic parts (aluminum–zirconium) and one organic functional ligand. Therefore, compared with a coupling agent such as silane, aluminum–zirconium organic metal complex coupling agents have more reaction points with inorganic fillers and can enhance the reaction with the surface of inorganic filler or paint.

Ziroa-luminate series coupling agents launched by Rhône-Poulenc of France not only can react with fillers and pigments containing hydroxyl irreversibly on their surface, but also have good reactivity with metals such as iron, nickel, copper, aluminum, and so on.[25]

5. *The borate coupling agent.* With boron atoms as the center, boric ester-type coupling agents can be used for surface modification of inorganic fillers containing boron. They can improve the compatibility between inorganic fillers containing boron and polymer. With a boron–oxygen skeleton, boric ester-type coupling agents can produce a good physical adsorption with a borate whisker. Therefore, they have good modification effects on modifying borate whiskers.

   Boric acid ester as a coupling agent was developed in the 1980s, and related research in China occurred later. Because boric acid ester is easy to hydrolyze, the application of borate coupling agents was restricted.[26] Triisopropyl borate ester was synthesized from boric acid by Chen et al.[27] in 1996, followed by isopropyl-2 (dodecyl) boric acid ester and isopropyl-2 (octadecyl) boric acid ester. Although the hydrolysis of boric acid esters was inhibited and stability was improved, such boric acid esters only have long-chain alkyl and hydroxyl yield by hydrolysis, and do not have reactive groups that can react with polymers. Therefore, they are not real coupling agents. Triethanolamine borate ester (BE_3) was synthesized from boric acid by Hu and Liang[28] in 2004, and the hydrolytic stability was greatly improved. Butoxyl borate esters with different degrees of branching were synthesized by Qinghai Institute of Salt Lakes, Chinese Academy of Sciences (shown in Table 2.3). A butoxyl with a higher degree of branching can inhibit water nucleophilic attack on boron atoms, thus enhancing the hydrolysis stability of boric acid ester. Among them, *tert*-butyl alcohol-diethanolamine borate ester, with the highest degree branching and greatest steric hindrance, has the best hydrolysis stability.[26]

6. *Rare earth coupling agent.* A rare earth compound is a material with a unique performance and is considered to have the capability to construct a treasure trove of objects

Table 2.3 Structures of Boric Acid Ester with Different Degrees of Branching

Boric Acid Ester	Structures
BE3	$H_2NH_2CH_2CO$—B(—$OCH_2CH_2NH_2$)—$OCH_2CH_2NH_2$
n-BE4	$H_2NH_2CH_2CO$—B(—$OCH_2CH_2NH_2$)—$OCH_2CH_2CH_2CH_2$
sec-BE4	$H_2NH_2CH_2CO$—B(—$OCH_2CH_2NH_2$)—$OCHCH_2CH_3$—CH_3
tert-BE4	$H_2NH_2CH_2CO$—B(—$OCH_2CH_2NH_2$)—$OC(CH_3)(CH_3)CH_3$

in the information era. Unexpected unique properties often appear when a small amount of rare earth is added to a system. Therefore, rare earth compounds are called "industrial monosodium glutamate."

Rare earth coupling agents have the following functions. (1) They change the surface properties of inorganic particles and the viscosity of composite resin to improve the compatibility and dispersion of inorganic particles in resin matrix, and produce interfacial interaction between the inorganic particles and matrix resin. (2) Inorganic particles modified by rare earth coupling agents can promote plasticization and internal plasticization, improve thermal stability and lubricity, and have a good synergistic effect with each composition of mixing resin. (3) They have good synergistic effects with traditional coupling agents, surface treatments, and activated inorganic powder filler. They not only can promote the dispersion of filler in the matrix resin, but can also improve the interface bonding between the packing and matrix or between the matrix and coupling agent. They overcome the shortcomings of the weak reaction of the traditional coupling agent with the matrix and can serve as a compatibilizer.[29–31]

In recent years, calcium carbonate treated with new rare earth coupling agents was applied to a PVC system. The results show that the dispersity of calcium carbonate in PVC and the compatibility with PVC improved, and mechanical properties and processability of the system were significantly improved and enhanced.[32]

In addition to the aforementioned common coupling agents, there are other coupling agents, such as aluminum–titanium composite coupling agent, zirconium coupling agents, and so on.

At present, the development and application of coupling agents are still progressing rapidly, and the main direction is to look for new types of coupling agents that are less costly and more efficient to help solve the problems of processing and mechanical properties of composite materials under high filling content.

2.3.1.1.2 Surface Grafting Modification

Surface graft modification includes direct graft polymerization modification and surface-initiated grafting polymerization modification. The former requires the particle surface to have active groups to copolymerize with other monomers, whereas the latter require that active groups to copolymerize with other monomers are generated from the grain surface by chemical or physical methods.

Carbon black has a strong ability to capture free radicals, so monomers can be directly polymerized on its surface. For inorganic particles with no reactive species, the initiator can be introduced onto the surface of inorganic ultrafine particles to form a monolayer, and then the initiator monolayer induces controllable "active" free radical polymerization. Liu connected the epoxy groups to the surface of silicone balls by the condensation reaction between the coupling agent KH560 and the hydroxyl groups on a nano silica surface.[33] Then through the ring opening reaction between the carboxyl group and epoxy groups of the reversible addition-fragmentation chain transfer (RAFT) reagent, the dithiobenzoate key was grafted to the surface of silicon ball particles, and the alternating copolymer of styrene and maleic anhydride was grafted onto the surface of the silicon ball.

2.3.1.1.3 Modification by Reaction with Fatty Acid or Alcohol

As a result of their large polarity, the surfaces of inorganic filler always adsorb hydroxyl or water molecules, which can be subjected to an esterification reaction with the carboxyl of fatty acids or alcohol carboxyl. Consequently, an organic layer is introduced on the surface of inorganic particles. Therefore, the surface energy of the inorganic filler decreases and the hydrophobicity increases. For example, calcium carbonate was treated by stearic acid. The chemical reactions on the surface of calcium carbonate are shown in Figure 2.3.[34]

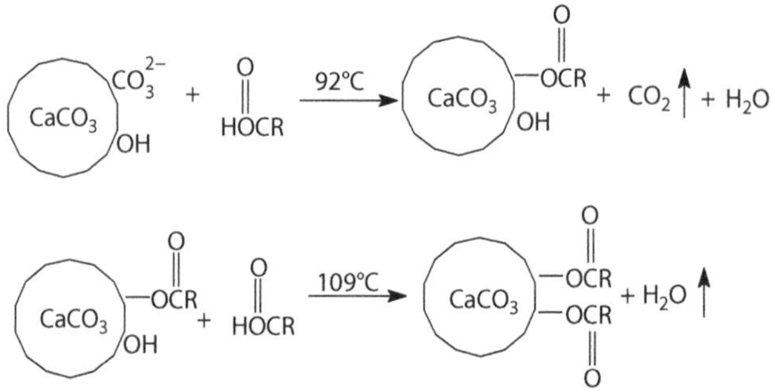

Figure 2.3   Chemical reaction on the surface of calcium carbonate.

2.3.1.2 Surface Coating Modification

In surface coating modification the coating material is enshrouded uniformly to objects to form a continuous complete coating using physical or chemical adsorption principles. This method is suitable for almost all kinds of surface modifications of inorganic fillers. The surface coating modification agents are generally inorganic or organic compounds. These materials are connected by physical means or Van der Waals forces and are selectively adsorbed on the surface of the packing, thereby weakening the agglomeration between particles. The modifiers used are surfactant, superdispersant, inorganic matter, and so on.

2.3.1.2.1 The Surfactant

A surfactant is a material that has fixed hydrophilic and lipophilic groups and can arrange directionally and absorb on the surface of a solution and reduce the surface tension of the solution significantly. Surfactants can be divided into anionic, cationic, nonionic, and amphoteric ions. The most common anionic surfactants are senior fatty acids and their salts, and the most common cationic surfactants are senior amines and their salt. The nonionic surfactants are silicone oil, fatty alcohol polyoxyethylene ether, fatty acid polyoxyethylene ether, and so forth.

Surfactants are generally composed of hydrophilic groups and hydrophobic groups. Hydrophilic groups with high polarity adsorb on the surface of inorganic particles, and hydrophobic groups with low polarity reach the solvent to create an obstruction. The dispersion of the particles into an organic solvent is improved. Bilayers are formed on the surface of the particles by surfactants when the concentration of surfactants is greater than the critical micelle concentration (CMC). The particles are hydrophilic, so the concentration of surfactant should be strictly controlled. The most common surfactants for filler surface modification are senior fatty acid and its salts, such as sodium stearate, calcium stearate, senior amine salt, and nonionic surfactant.

2.3.1.2.2 The Hyperdispersant

The hyperdispersant is a special kind of surfactant that is similar to the traditional surfactant in its amphiphilic structure. But instead of hydrophilic groups and hydrophobic groups of the surfactant, the hyperdispersant has an anchoring group and solvation chain. The most common anchoring groups are $-NR_2$, $-N^+R_3$, $-COOH$, $-SO_3H$, multiple amines, polyols, polyethers, and so on, which can be adsorbed tightly on the surface of solid particles through ionic bonds, covalent bonds, hydrogen bonds, and Van der Waals forces, thereby preventing hyperdispersant stripping. The most common solvation chains are polyesters, polyethers, polyolefins, polyacrylates, and so forth. According to the polarity, solvation chains can be divided into three types: polyolefin chains with low polarity, polyester or polyacrylate chains with medium polarity, and polyether chains with strong polarity. In the polarity matching dispersion medium, the solvation chain has good compatibility with the dispersion medium, and a protective layer with sufficient thickness is formed on the surface of solid particles. A hyperdispersant styrene-maleic anhydride-butyl (SMB) methacrylate copolymer with maleic anhydride and its single ester content as anchor groups, butyl methacrylate (BMA) as the solvation chain, and styrene as functional groups was prepared via a radical polymerization method by Li et al.[35] The result shows that the modification effect of nano TiO_2 powder by using SMB-2 is better when the concentration of SMB-2 is 8%. In addition, compared to the traditional titanate and silane coupling agents, the modification effect of hyperdispersant is better.

The choice of hyperdispersant needs to consider two aspects. (1) The first is the polarity of the dispersion medium and the solvent solubility to the solvation chain of hyperdispersant. It is generally hoped that the dispersion medium has a strong ability to dissolve the solvation chain and a weak ability to dissolve anchor groups. (2) The second is the surface properties of particles to be dispersed, such as surface polarity, surface functional groups, surface acid–base property, and so on. Particles with low polarity require a hyperdispersant containing multipoint anchoring groups. In addition, different functional groups have different reactivities and modes of action; different adsorption positions of the particle surface will also selectively adsorb the anchor groups depending on different pH values. Generally, a hyperdispersant is a mixture of general multipolarity groups. So the same hyperdispersant has a good dispersion effect for particles with different surface properties.

2.3.1.2.3 Inorganic Substances

Until now, there have been few reports on inorganic particles modified by inorganic substances, but this study has begun to attract attention. For example, the calcium carbonate whiskers and potassium titanate whiskers that are coated with an inorganic silicon dioxide layer possess a silica nature to some extent; coupling agents therefore can bond with the silica on the surface which is advantageous for whisker surface modification. In addition, surface smoothness, whiteness, acid resistance, dispersibility, specific surface area, and so forth of whiskers can increase significantly, thus greatly improving the applicability of whiskers.[36]

2.3.1.3 Other Modification Methods

2.3.1.3.1 Mechano-Activated Modifying Method

Mechano-activated modification uses a mechanical method, such as crushing, grinding, and friction, to change the lattice structure and crystal structure of the packing, increase system internal energy and temperature, promote particle melting and thermal decomposition, produce free radicals or ions, reinforce surface activity of fillers, promote chemical reactions between packing and other material or attachment to each other, and therefore achieve the aim of surface modification.

2.3.1.3.2 Microencapsulation Modification

Microencapsulation modification consists of coating the surface of particles with an even membrane of other substances that has a certain thickness and changes the surface characteristics of particles. The role of the capsule is to control the release of the isolated and hidden material. Microcapsule-reinforced inorganic particles were synthesized through *in situ* polymerization on the surface of inorganic particles by Chang et al.,[37] and the microcapsule-reinforced inorganic particles have good compatibility with the polymer matrix.

In addition, ultraviolet light, infrared rays, corona discharge, and a plasma radiation halving method can also be carried out on particle surface modification. These methods do not need modifying agents, and are therefore beneficial for environmental protection. However, these technologies are complex and expensive.

2.3.2 Surface Modification Technique of Inorganic Fillers

The surface modification technique of inorganic fillers directly affects the filling modification.[7,8] The common surface modification methods can be classified into three types: inorganic filler pretreatment, blending, and masterbatch methods.

2.3.2.1 Pretreatment Method

In the pretreatment method, fillers are surface modified before mixing with a resin matrix. Pretreatment methods can be divided into two kinds: dry processing and wet processing.

In dry processing, surface modification agents are directly sprayed to the powder surface, then fully mixed, dried, or subjected to other treatments. Advantages of dry processing are its simple operation, the fact that the content of modification agents can be accurately controlled, and so on.

In wet processing, treatment agents are first dispersed in aqueous solution; then fillers are added and the mixture is stirred so the molecules of the modification agent are adsorbed to the surface of fillers through surface adsorption or chemical bonds, and finally filtered and dried. Wet processing can coat uniformly, but has a cumbersome operation process, low efficiency, and high cost.

The pretreatment method can play a good role as a coupling agent and is more economical.

2.3.2.2 Blending Method

In the blending method, resin, filler, surface modification agent, and so on are directly added to the reactor and blended together. It has some overlaps with dry processing, but the blending method is simple and does not require extra equipment. Compared to the pretreatment method, a higher concentration of the coupling agent (two to three times) is required to achieve the same effect.

2.3.2.3 Masterbatch Method

A masterbatch is made by mixing, extruding, and granulating filler, modifying agent, additives, and carrier resin by a mixer or twin-screw extruder, then adding these to the resin for injection molding or molding. The performance of the product prepared using the masterbatch method is stable. The masterbatch method is also the main form of application of fillers.

2.4 Interface between Inorganic Fillers and Polymer

2.4.1 Formation and Structure of the Interface

Generally, the polymer-based composite is a composite of a polymer with a relatively low modulus and an inorganic filler with a relatively high modulus. In the structure, there are significant differences between most of the matrix resins and the inorganic filler, and they are incompatible in chemical and physical properties. So there is an interface between them. To a large extent, the performance of the composite material depends on the combination of inorganic and organic phases in the composite material.

The formation of the interface generally has two stages. The first stage is the contact and wetting process of the substrate with the filler; the second stage is the formation of the fixed interface in the process of polymer curing physical or chemical changes. The second stage is affected by the first stage. At the same time, the second stage affects the structure of the interface directly.

The interface between the filler and the polymer is a complex interface layer. Although the interface layer is very thin (only a few molecules in thickness), it possesses a tremendously complex structure, and the chemical composition, mechanical properties, molecular arrangement, and thermal performance of the interface layer change continuously, as shown in Figure 2.4.

The filler and the matrix form a whole by the interface. Stress can be transferred evenly through the interface, and the area shows excellent performance. In addition, as a result of the existence of the interface, composite materials produce discontinuous physical properties, such as interface friction and electrical resistance, heat resistance, dimensional stability, impact resistance, and so forth, and can prevent the expansion of the crack and reduce stress concentration. Therefore,

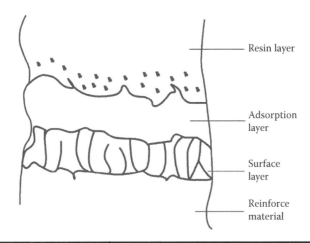

Resin layer

Adsorption layer

Surface layer

Reinforce material

Figure 2.4   Interface model of composite materials.

the formation of a complete interface area is required in the preparation process of composite materials.

The interface bonding force exists on the interface between two phases, so interface strength is created and a composite effect is produced. The interfacial binding force has macroscopic and microscopic aspects. The macroscopic binding force refers mainly to geometric factors, such as mechanical hinge force, uneven surface, cracks, and voids; the microscopic binding force includes chemical and valence bonds. A chemical bond is the strongest binding force, usually formed by an interface chemical reaction. To increase the interface bonding force between the filler and matrix, the surface of inorganic fillers is usually treated.[7,8,38]

2.4.2 Interfacial Mechanism

The interfacial mechanism refers to the microcosmic mechanism of the interface function. Generally, the resin phase is in solution or melt flow state to contact with the filler to form composite materials after a curing reaction or cooling curing. In this process, how the resin interacts with the filler becomes of more and more concern.

A variety of interface function theories of composite materials have been proposed, but none is completely adequate. The following are current relatively popular and important theories.

2.4.2.1 Infiltration Theory

The infiltration theory was put forward by Zisman in 1963. This theory holds that infiltration is one of the basic conditions of forming an interface. If two ideal clean surfaces combine by physical means, the resin and the filler surface must have good infiltration to make them contact closely. When both have very good infiltration, the liquid phase can be extended to the concave pit of another phase surface. Thus, two phases have a large contact area and combine closely. The bonding strength of the phase with high surface energy provided by physical adsorption would be more than the cohesive energy of the resin, so that the two phase objects have good bonding strength.

However, to bring about complete infiltration of the surface, the surface tension of the resin must be lower than the critical surface tension of the filler. Surface modification agents used in polymer matrix composites change the surface state of fillers so the fillers will have good infiltration with the resin, and as a result the filler and the resin will have good bonding strength.

The good wettability between the polymer matrix and the filler will help improve the interface of the composite strength, but infiltration is not the only condition of the interfacial bond. For example, the surface tension of vinyl silane is 33.4 mN/m, and it is an effective coupling agent for an unsaturated polyester; the surface tension of ethyl silane is similar to that of vinyl silane, but it is not valid for an unsaturated polyester.[38]

2.4.2.2 Chemical Bonding Theory

Chemical bonding theory is a widely used and successful theory of interface interaction. This theory holds that the matrix resin should have active groups that can react with the surface of fillers to realize effective bonds between two phases. Through reactive chemical reaction of active groups, an interface between two phases is formed by chemical bonds and has good bonding strength. If two phases cannot directly chemically react with each other, they can form an interface by chemical bonds through a medium, such as a coupling agent, to improve the bond strength between the reinforcing material and the substrate, as shown in Figures 2.5 and 2.6.

Different coupling agents are chosen for different filling compound systems. As an example, the silane coupling agent

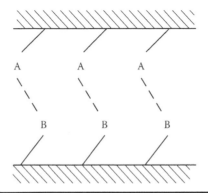

Figure 2.5   **Chemical reaction between two phases.**

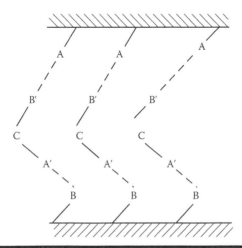

Figure 2.6   **The interface of two phases combined by chemical bonds through a coupling agent.**

an effective additive that can improve the interface bonding, was developed on the basis of the chemical bonding theory.

The chemical bonding theory has some limitations. For example, it is difficult to explain the thermal stress relaxation phenomenon of the composite material during the resin curing process. In addition, some coupling agents do not contain active groups that can react with the matrix resin, but they can interact through molecular long chains and resin molecular chains and therefore have a better treatment effect.

2.4.2.3 Interfacial Acid–Base Function Theory

The interfacial acid–base function theory was put forward the first time in 1962 by Fowks. This theory holds that the filler and polymer matrix of a composite material can be treated as a generalized acid or alkali. When the surfaces of the filler and polymer matrix have different acids or alkalis, they are easy to combine. The greater their difference in pH, the more easily they will combine. When the surfaces of the filler and polymer matrix have the same acid or alkali, the surface of the filler should be modified to change its pH value. The different surface pH values are beneficial to improve the bonding strength between the reinforced material and polymer matrix and improve the performance of filled plastics.

Acid parameters E and alkaline parameters C can be used to show the acid and alkaline values of the surface of materials. To measure the E and C values of material surfaces, R. S. Drago proposed the famous Formula 2.2:

$$\Delta H = E_A E_B + C_A C_B \tag{2.2}$$

where
  ΔH = the enthalpy change of interaction between two substances A and B
  E = the acid parameter
  C = the alkaline parameter

ΔH can be measured by inverse gas chromatography, as follows. The solid analyte is installed in gas chromatography as a solid stationary phase, a solvent with known values of C and E is used as a probe, and specific retention volume V_g^0 is measured at different temperatures. Then Formula 2.3 is used to calculate the enthalpy change of the interaction with the stationary phase.

$$\frac{\partial(\ln V_g^0)}{\partial\left(\frac{1}{T}\right)} = -\left(\frac{\Delta H - \Delta H_v}{R}\right) \tag{2.3}$$

where
ΔH_v = the vaporization heat of probe
ΔH = the enthalpy change of interaction between stationary phase and mobile phase
R = the gas constant
V_g^0 = the specific retention volume

If two ΔH values between the stationary phase and two probes with known C and E values are measured respectively, by solving simultaneous equations of Formula 2.2 and calculating acid–base parameters of the surface of the stationary phase, the acidity and alkalinity of the material surface can be judged.

Acid–base parameters of some solvents are shown in Table 2.4. Among them, chloroform has a smaller alkaline parameter C and larger acid parameter E, so it is acidic.

Table 2.4 Acid–Base Parameters of Some Solvents (C, E)

Solvent	Chloroform	tert-Butanol	Tetrahydrofuran	Dioxane
C	0.308	0.615	8.75	4.88
E	6.79	4.18	2.00	2.33

Table 2.5 Acid–Base Parameters of Some Materials (C, E)

Materials	PMMA	SiO_2	$\alpha\text{-}Fe_2O_3$
C	1.03	1.12	1.1
E	0.68	4.56	0.86

Acid–base parameters of some materials are shown in Table 2.5. The surface of polymethyl methacrylate (PMMA) is alkaline, the surface of SiO_2 is acidic, and the surface of $\alpha\text{-}Fe_2O_3$ is alkaline.

According to the measured C and E values of the surfaces of the filler and matrix, the interface bonding strength can be estimated. For example, when SiO_2 and $\alpha\text{-}Fe_2O_3$ are used to reinforce PMMA respectively, the strength of the interface adhesion SiO_2/PMMA is larger than that of $\alpha\text{-}Fe_2O_3$/PMMA.

The C and E values of SiO_2/PMMA and $\alpha\text{-}Fe_2O_3$/PMMA are substituted into Formula 2.2, respectively. Both interaction enthalpy changes can be obtained: ΔH (SiO_2/PMMA) = 17.85 kJ/mol and ΔH ($\alpha\text{-}Fe_2O_3$/PMMA) = 7.21 kJ/mol. Thus, it is estimated that the interface bonding strength of SiO_2/PMMA is greater than that of $\alpha\text{-}Fe_2O_3$/PMMA. This result is the same with the forecast estimation of acid–base parameters.[38]

2.4.2.4 Transition Layer Theory

The transition layer theory holds that additional stress would be generated on the interface between the filler and the matrix during molding of composites because of different expansion coefficients of the filler and the matrix. In addition, under the action of external loading, the uneven distribution of stress in the composite material can produce a stress concentration phenomenon in some parts of the interface. Therefore, there is a transition layer in the interface area that plays an important role in stress relaxation.

Two theories were put forward by Kinloch and Kodokion et al. on the formation of the interface and its mechanisms. One holds that the treating agent forms a layer of plastic on the interface, and plastic deformation can relax or reduce the interfacial stress. This theory is called the deformation layer theory. However, the theory cannot explain the traditional interface processing method; that is, the amount of coupling agent on the surface is not enough to meet the requirement of stress relaxation. Therefore, the first adsorption theory and flexible layer theory are proposed on the basis of this theory. The plastic layer is formed by the coupling agent and the flexible layer by preferential adsorption. Through the deformation of relaxation stress of the flexible layer, the development of crack and the interface bonding strength are improved. Flexible layer thickness has nothing to do with the number of coupling agents themselves in the interfacial area. Grafting a flexible rubber layer on the reinforcement fiber surface is the typical application case of the deformation layer theory.

Another view is the inhibition layer theory, which holds that the structure of the stress relaxation transition layer between the matrix and the packing is not a flexible deformation layer, but an interfacial layer with a modulus between that of the matrix and the packing. The treatment agent is a part of the interfacial area, and this part is a substance with a medium modulus that is between that of the high-modulus reinforcing material and the low-modulus matrix. It has the effect of transferring stress uniformly and reducing interfacial stress, so that it can enhance the performance of the composite materials. The inhibition layer theory is not widely accepted because it lacks the necessary experimental basis.[32]

2.4.2.5 Friction Theory

Friction theory holds that the formation of the interface between the polymer matrix and the filler is due to friction, and the friction factor between the matrix and the packing determines the strength of the composite material. The treatment agent can increase the friction factor between the substrate and the packing and therefore increase the strength of the composite material.

The theory can better explain how the strength of the composite interface declines after immersion in low molecular substances such as water, and the strength will partially recover after drying. The friction factor between the substrate and the packing decreases when small molecules such as water immerse in the interface, so the stress transfer ability of the interface increases and the strength decreases. When the interface has less moisture after drying, the friction factor between the substrate and the packing increases, so the stress transfer ability of the interface will increase and the strength will partially recover.

2.4.2.6 Diffusion Theory

The diffusion theory was first proposed by Borozncui and holds that the mutual cohesion between polymers is caused by mutual diffusion of large molecules on the surface. The diffusion, osmosis, and association of molecular chains of two phases form the interfacial layer.

The diffusion of the interfacial system is similar to the dissolving process of substances. They are all hybrid processes. The diffusion effect leads to the interface becoming fuzzy and even disappearing (e.g., solid dissolved in liquid). The diffusion process correlates with molecular weight, flexibility, temperature, solvent, plasticizer, and other factors of the molecular chain. The interface diffusion between the polymer matrix and the filler could improve bonding performance. However, the diffusion theory cannot explain the adhesion phenomenon between the polymer matrix and inorganic reinforced material without interfacial diffusion.

Researchers studying the interface interaction mechanism also put forward other theories of interface interaction, such as electrostatic theory, theory of reversible hydrolysis, the theory of adsorption, and so on. Various theories explain the interfacial interaction from different angles, but no single theory can perfectly explain all kinds of interfacial phenomena. With the

development of polymer science and the interface characterization technique, the interface theory will be further developed and perfected.

2.4.3 Interfacial Damage

There are many factors that can damage the interface of
composite materials, such as mechanical, thermal, and optics
factors, and so on. Previously, the macro-debonding phenomenon during damage of a composite material was given
more attention. In recent years, the concept of the destruction
of the resin interface layer was proposed based on micro-
noninterface debonding. Therefore, there are two forms of
damage of composite materials.

2.4.3.1 Matrix Resin Damage

When the interface bonding ability of the composite material
is strong, the strength of the filler is high, and the intensity of
the resin matrix is relatively low, matrix resin damage is prone
to occur. High strength and high elongation of the resin matrix
and good bonding interface can greatly improve the strength
of the filled polymers.

2.4.3.2 Interfacial Damage

If the interfacial bonding strength of the composite material
is less than the cohesive strength of the matrix resin and the
strength of the filler, a stress concentration near the filler can
easily occur when composite materials are subjected to shear
or tensile stress. Debonding damage will occur near the interface if the interfacial bonding is weak.

In short, when there is an external force against composite materials, in addition to the filler and matrix under stress,
the interface plays a very important role. The modification
of the interface between two phases to form an appropriate

bonding force and a modulus layer with an interaction match that can smoothly transfer stress in the middle of the modulus to improve the mechanical properties of composite materials is always an important research field of polymer material science.

2.4.4 Characterization of the Interface

In recent years, the characterization technology of resin matrix–filler interface interaction has undergone many new and important developments. This laid the experimental foundation for studies of interfacial composition and structural form, speculation of many physical and chemical properties of composites, the interfacial interaction mechanism, interface forms, and interface optimization design. The following are methods commonly used for characterizing the interface.

2.4.4.1 Contact Angle Method

The contact angle method is a thermodynamic method in which liquid (such as liquid resin, liquid coupling agent, etc.) and solid filler surface contact to achieve balance. The infiltration theory shows that good infiltration status is the premise of the strong interaction. T. Young put forward the famous Young's Equation 2.4 from research on force equilibrium conditions when a general liquid is an adhesive on a solid surface.

$$\gamma_l \cos \theta = \gamma_s - \gamma_{sl} \qquad (2.4)$$

where
  γ_l = the surface tension of the chosen liquid
  γ_s = the surface tension of the solid (polymer)
  γ_{sl} = the liquid–solid surface tension

Young's equation is the basis of the study of liquid–solid infiltration function, and the contact angle is often used to measure the infiltration effect of the liquid on the solid. The surface tension data can also represent the strength of the interfacial effect. Through regulating the ratio among three kinds of surface tension, the infiltration condition of the interface and the interfacial effect can be improved.

2.4.4.2 Viscosity Method

The viscosity method is used to characterize the affinity of the filler and the polymer based on the rheological behavior of a filled polymer system.

It is found that when compared with the corresponding pure polymer system, the melt viscosity of the filled polymer system at a high shear stress range reaches a stable value.

Different filled polymer systems with the same kind and same volume content of fillers have different viscosity values because of the different thicknesses of adsorption layer adsorbed on the filler surface. The thicker the adsorption layer, the more stable is the viscosity and the stronger the interfacial interaction.

2.4.4.3 Mechanical Strength Method

Determination of the macromechanical strength of composite materials can characterize whether the micropolymer and the filler interface reach an ideal state, and this method is common.

2.4.4.4 Microscopy Method

The microscopy method is a method in which the interfacial condition between the fillers and polymer matrix can be directly observed. The most commonly used is scanning electron microscopy (SEM). By SEM observation of the morphological structure of impact or tensile cross-sections of polymer matrix composites, the filler dispersion in the overall situation and overall interfacial adhesion of the fillers and polymer can be analyzed. If the surface of the filler is clear and with less resin adhesion, the interfacial bonding is poor.

In recent years, microscope technology has undergone new developments, such as transmission electron microscopy (TEM), atomic force microscopy (AFM), scanning tunneling microscopy (STM), and so forth.

2.4.4.5 Thermal Analysis

The purpose of thermal analysis is to determine changes in the nature of the sample with time or temperature in a specified atmosphere under controlled temperature conditions. The main thermal analyses are differential thermal analysis (DTA), differential scanning calorimetry (DSC), dynamic thermal analysis (DMA), thermogravimetry (TG), thermal mechanical analysis (TMA), and so on. Among them, the thermogravimetric method is commonly used. If the interfacial adhesion is strong, the motion of the polymer chain segments is difficult near the interfacial area and the thermal weight loss temperature and heat resistance increase. These can reflect the interfacial performance information.

In addition to the aforementioned research methods, there are other methods, such as Fourier transform infrared spectrometry (FTIR), X-ray photoelectron spectroscopy (XPS), the secondary ion mass spectrometry (SIMS), ion-scattering spectroscopy (ISS), combined methods such as XPS–SIMS, and so forth.

Each characterization technique has its own features. According to the purposes of characterization, one needs to choose a proper method for different polymers to achieve desired effects. With the rapid development of modern instruments, there will be more analysis methods for interface characterization of polymer matrix composites which will lead to clearer understanding of the structure and properties of polymers.

2.5 Analysis of Filling Compound Effects

The polymer shifts to a composite material from a single material after filling, and composite effects on performance are produced. Compound effects can be divided into the following three kinds.[39]

2.5.1 Effect of the Component

Under the condition of known physical and mechanical properties of components, the effect of the component considers only the volume fraction and weight fractions of the fillers as variables regardless of the influence of shape, orientation, size, and other variables of the components.

The relationship between the properties of a two-component system and performances of components could be estimated by the mixture rule. The two most commonly used relations are as follows:

$$P = P_1\beta_1 + P_2\beta_2 \tag{2.5}$$

$$\frac{1}{P} = \frac{\beta_1}{P_1} + \frac{\beta_2}{P_2} \tag{2.6}$$

where
 P = specific properties of a two-component system, such as density, electric properties, modulus, and so forth
 P_1, P_2 = corresponding performance of components 1 and 2
 β_1, β_2 = mass fraction or volume fraction of components 1 and 2

2.5.2 Effect of the Structure

The structure effect refers to the fact that when compound properties are considered as component properties and composition functions, the structure morphology of the continuous phase and the dispersed phase, and the orientation and size of the dispersed phase must be considered. The performance of composite materials cannot be estimated using the aforementioned mixture rule. In most cases, Formula 2.5 can only give the upper limit of compound performance, while Formula 2.6 can give the lower limit. Taking elasticity modulus as an example, Formula 2.5 applies to the elasticity modulus of the composite when the continuous phase has a large modulus and the dispersed phase has a small modulus. Formula 2.6 applies to the modulus of elasticity of the composite when the continuous phase has a small modulus and the dispersed phase has a large modulus. In Figure 2.7, curve 2 shows the measured values of the composite.

In area AB, the continuous phase has a small modulus, and the measured values have good congruity with theoretical curve 1 according to Formula 2.6; in area CD, the continuous phase has a large modulus, and the measured values have

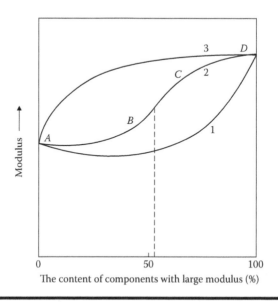

Figure 2.7 The relationship of modulus of elasticity of the composite with components. Curve 1, theoretical value; curve 2, measured value; curve 3, upper limit value.

good congruity with theoretical curve 3 according to Formula 2.5; area BC is the phase transition area of the composite.

2.5.3 *Effect of the Interface*

The interface effect is the main part of the composite effect. The performance of the interfacial area is different from that of each pure component, and it can be regarded as the third phase C excepting A and B, as shown in Figure 2.8.

If the existence of the interfacial area is not considered, and A and B phases are combined as volume fractions Φ_A and Φ_B, the addition law of X performance of composite material is:

$$X = \Phi_A X_A + \Phi_B X_B \tag{2.7}$$

Because the interfacial region has formed a new phase, assume the volume of the new phase is Φ_C; then Φ_A becomes Φ'_A and Φ_B becomes Φ'_B.

$$X = \Phi'_A X_A + \Phi'_B X_B + \Phi'_C X_C \tag{2.8}$$

If A and B are mixed equally:

$$X = \Phi_A X_A + \Phi_B X_B + \Phi_C \Delta X_C \tag{2.9}$$

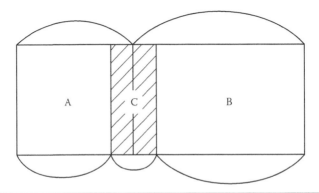

Figure 2.8   Formation of interface phase C.

where

$$\Delta X_C = X_C - \frac{X_A + X_B}{2}$$

Formula 2.8 can be switched to

$$X = \Phi_A X_A + \Phi_B X_B + K\Phi_A X_B \qquad (2.10)$$

Formula 2.10 is referred to as the second law of composites, where K is related to phase C, which is related to ΔX_C, and K is known as the interaction parameter of A and B.

The performance of the compound changes over Φ_B, and is shown in Figure 2.9.

From Figure 2.9 we can see that one curve has a maximum value when $K > 0$, and one curve has a minimum value when $K < 0$. That is, for X to have a maximum value, an interface area must be formed and the property of this interface must be higher than the arithmetic mean of the original components.

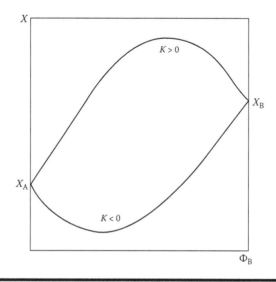

Figure 2.9   Second law of composites.

Take the calculation of the elastic modulus of fiber-filled composite material as an example. Most of the fibers are anisotropic, so different orientations of fibers lead to different properties of the composite. For a fiber composite material with uniaxial orientation, fibers are arranged in the same direction, and there are four important moduli: the longitudinal Young's modulus E_L, where the load acts along the direction of fiber orientation; the transverse Young's modulus E_T, where the load acts along the direction perpendicular to the fibers; the longitudinal shear modulus G_{LT}, where shear stress acts along the fiber direction; and the transverse shear modulus G_{TT}, where the shear stress is perpendicular to the fiber direction.

When the fiber is very long, the longitudinal Young's modulus (E_L) can be calculated by the following formula:

$$E_L = E_1 \Phi_1 + E_2 \Phi_2 \tag{2.11}$$

where E_1, E_2 and Φ_1, Φ_2 are the modulus and volume fraction of the matrix and fiber packing respectively.

The modulus of the composite is commonly shown by specific modulus (M/M_1), which is the ratio of the modulus of the composite and matrix. The specific modulus can be calculated by the following formula:

$$\frac{M}{M_1} = \frac{1 + AB\Phi_2}{1 - B\psi\Phi_2} \tag{2.12}$$

where
M = Young's modulus, shear modulus, or bulk modulus
A = a constant that is introduced considering the geometry of the enhancer and the Poisson ratio of the matrix
B = a constant that is associated with the modulus ratio of the reinforcing agent and matrix, $B = \dfrac{M_2/M_1 - 1}{M_2/M_1 - 1 + A}$

Ψ = a value associated with the maximum packing coefficient Φ_m of the reinforcing agent, $\psi \cong 1 + \left(\dfrac{1-\Phi_m}{\Phi_m^2}\right)\Phi_2$, with the fiber reinforcing agent generally taking a value of $\Phi_m = 0.8–0.9$

The A values of different systems are listed in Table 2.6; the relationships between different specific moduli of composite material and fiber content are shown in Figure 2.10.

Formula 2.11 applies only to the long fiber. When the fiber is relatively short, modulus E_L is smaller than the value calculated according to Formula 2.11. Figure 2.11 shows the relationship diagram of E_L/E_1 and the fiber length-to-diameter ratio L/D.

In general, to obtain the highest strength and modulus, the L/D must be at least 100.

If the load is rotated 90°, the modulus will change enormously and E_L will become E_T. The influence of the angle θ between the load and fiber on the modulus E_θ is shown in Figure 2.12.

Table 2.6 A Values of Fiber-Filled Composites

Composite Material Type	Modulus	A	Composite Material Type	Modulus	A
Longitudinal (uniaxial orientation)	E_L	$2L/D$	Random orientation (3D)	G with $L/D = 4$	2.08
Transverse (uniaxial orientation)	E_T	0.5	Random orientation (3D)	G with $L/D = 8$	3.80
Uniaxial orientation	G_{LT}	1.0	Random orientation (3D)	G with $L/D = 15$	8.38
Uniaxial orientation	G_{TT}	0.5	Random orientation (3D)	G with $L/D = \infty$	∞

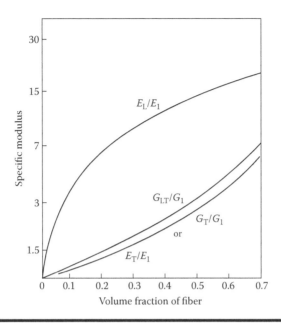

Figure 2.10 The relationship between the specific modulus of uniaxially oriented glass fiber-reinforced epoxy resin and fiber content.

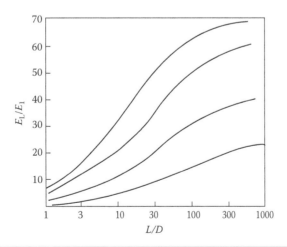

Figure 2.11 The relationship between E_L/E_1 and fiber length-to-diameter ratio *L/D*.

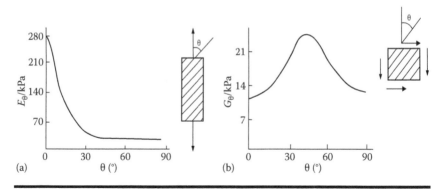

Figure 2.12   The relationship between the specific modulus of boron fiber-epoxy resin composite and θ. (a) Vertical loading. (b) Transverse load.

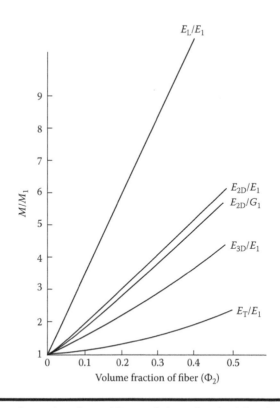

Figure 2.13   The ratio of specific modulus of uniaxial orientation composite material to biaxial and three-dimensional orientation composite material (E_2/E_1 = 25).

Uniaxially orientated fiber composite material has a high modulus in only one direction. To obtain composite materials with good mechanical properties in at least two or three directions, the fiber should have random orientation. Figure 2.13 shows the Young's moduli E_{2D} and E_{3D} and shear moduli G_{2D} and G_{3D} of the composite material, which has an in-plate random orientation and three-dimensional random orientation.

As shown in Figure 2.13, the moduli of fiber composite materials in a random orientation are generally larger than those of the matrix, but are much lower than E_L.

Thus, inorganic whiskers as fillers have the anisotropy of fibers and could support the skeleton function. In addition, inorganic whiskers could be evenly dispersed in the composite material because of its small size and would not lead to anisotropy. Therefore, polymers filled with inorganic whiskers are more advantageous.

References

1. Fuzhi Wang. Development and application of inorganic filler modified. *Plastics Technology*, 25(4):8–12, 1988.
2. Yingjun Liu, Boyuan Liu. *Plastic's filling and modifying*. Beijing: China Light Industry Press, 1998.
3. Jianping Dong, Xianlan Ji, Yanwu Lei. Cost analysis of calcium carbonate filled plastic. *Chemical Industry*, 25(9):32–35, 2007.
4. Yingjun Liu. Talc powder in the plastic modification effect and application prospect. *China Non-Metallic Mining Industry Herald*, 25(4):3–7, 2005.
5. Tongkao Xu. Plastic filling modified masterbatch production and market. *Plastic Processing*, 41(1):30–36, 2006.
6. Xueyong Zhou. *Polyethylene/calcium carbonate composite films developed degradation*. Wuhan: Huazhong Agricultural University, 2001.
7. Shen Wang. *Polymer modification technology*. Beijing: China Textile Press, 2007.
8. Jing Guo. *Modification of high polymer material*. Beijing: China Textile Press, 2009.

9. Yanming Dong. Packing modification processing. *Hunan Plastic*, (1):20–24, 1983.
10. Yunfang Zhao. The preparation and properties of glass fiber reinforced polyphenylene oxide research. *Chemistry and Adhesion*, 30(4):47–49, 2008.
11. Minghui Du, Xing Chen, Hui He et al. Study on property of MMT/PP/PVC blends. *New Chemical Materials*, 39(11):135–137, 2011.
12. Xuedong Li. Plastic toughening rigid filler. *Plastic*, 24(5):5–12, 1995.
13. Huixuan Zhang, Xiaoye Bao, Haidong Yang. Toughening mechanism of plastics. *Journal of Changchun University of Technology*, 23(S):77–79, 2002.
14. Xiaoguang Zhu, Xiaohua, Deng Xuan Hong et al. Toughening polypropylene composites filled fracture toughness and the toughening mechanism. *Acta Polymerica Sinica*, 39(2):195–201, 1996.
15. Yaguang Qi. Industrial progress of the conductive plastic in the world. *Engineering Plastics and Their Applications*, 36(3):73–77, 2008.
16. Huaiyuan Wang, Xin Feng, Yijun Shi et al. Tribological behavior of PEEK composites filled with various surface modified whiskers. *Journal of University of Science and Technology Beijing*, 29(2):182–185, 2007.
17. Chunshan Fang, Zhongyi Zhang, Rui Huang. The effect of some characteristics of fillers on filled plastics. *Plastic*, 16(3):29–34, 1987.
18. Deifa Shang. *Certain properties of filler on the properties of filled plastics*. Haerbin: Harbin University of Science and Technology, 2006.
19. Qunhua Xu, Wei Meng, Xujie Yang et al. Toughened and strengthened unsaturated polyester resin by nanometer titanium dioxide. *Polymer Materials Science and Engineering*, 17(2):158–160, 2001.
20. Guiying Li, Ming Liang, Qingshan Li. The application and the new progress of coupling agent. *Chemical Industry Times*, 13(3):21–25, 1999.
21. Zhen Zhao, Wenlong Zhang, Chen Yu. The research progress and application of the coupling agent. *Plastic Additive*, 10(3): 4–10, 2007.
22. Dongshen Mei, Wei Chen, Zaishan Tao et al. Titanate coupling agent and its application. *Plastic Additive*, 2(3):12–15, 1998.

23. Kairong Liao, Chenmou Zheng, Yongqian Qu. Synthesis of aluminate coupling agents and their modification on PVC/L-CaCO$_3$ and PVC/L-CaCO$_3$ blends. *Polymer Materials Science and Engineering*, 11(6):40–44, 1995.

24. Wengong Zhang, Lianhui Wang, Wenhua Zhang. Synthesis and space configuration of coupling agent—Coordination aluminate. *China Plastics*, 14(7):79–85, 2000.

25. Yuru Chen. The application of aluminum zirconium coupling agent. *Plastics Industry*, 29(6):44–46, 2001.

26. Jing Dai. *The synthesis and characterization of boric acid ester coupling agent and borate whisker modified polymer research.* Xining: Qinghai Institute of Salt Lakes, Chinese Academy of Sciences, 2006.

27. Honggang Chen, Suyun Xiang, Lianjin Weng et al. Synthesis of boric acid ester coupling agent. *Plastics Science and Technology*, 24(2):23–25, 1996.

28. Xiaolan Hu, Guozheng Liang. Synthesis and characterization of boric acid ester coupling agent. *Journal of Functional Polymers*, 16(2):179–183, 2003.

29. De Zheng. Research progress of the rare earth additives' application in plastic. *World Plastics*, 22(7):45–50, 2004.

30. Guanghua Liu. *Rare earth materials and application technology.* Beijing: Chemical Industry Press, 2005.

31. De Zheng, Zhuangli Li, Wenzhi Wang et al. Application of rare earth modified PVC processing aid. *Plastics*, 33(2):1–5, 2004.

32. Xiaojuan Wang, Lan Zhou. Application of rare earth composite in plastics modification. *Plastics*, 33(4):13–15, 2004.

33. Chunhua Liu. *Grafting polymers from the surface of silica nanoparticles via living radical polymerization and their applications.* Hefei: University of Science and Technology of China, 2009.

34. Xinjun Li, Jingfa Liao, Long Qu et al. Study on the surface modification of CaCO$_3$ with stearic acid by thermal analysis and the characterization of the surface properties. *Chemical Research and Application*, 9(5):478–481, 1997.

35. Rongfu Li, Duxin Li, Yannan Zhang. Synthesis and application effect of hyper-dispersants for modification of inorganic nano-powders. *Materials Science and Engineering of Powder Metallurgy*, 16(2):214–221, 2012.

36. Hailiang Li, Guoquan Wang, Xiaofei Zeng et al. Effects of SiO$_2$ encapsulated nano-meter CaCO$_3$ whisker and MMT on mechanical properties of PBT. *Plastics Industry*, 35(2):32–35, 2007.

37. Suqin Chang, Tingxiu Xie, Guisheng Yang. A reinforced inorganic particle microcapsules and their preparation methods. Patent No. CN:200610024839.4.
38. Lixin Shen, Peiqiong Sun. The progress of the polymer matrix composites interface theory. *Glass Fiber Reinforced Plastics*, 41(4):19–30, 2011.
39. Liucheng Zhang, Xiongwei Zhai, Huili Ding et al. *The base of polymer.* Beijing: Chemical Industry Press, 2006.

Chapter 3

New Materials—
Inorganic Whiskers

Filler-modified polymer composites produced concomitant with deeper research in filling modification of polymer materials are now widely used. Various advanced filling materials have been developed and applied in some areas. An example is the coal ash hollow glass microsphere, a kind of light hollow microsphere substance prepared from coal ash, which is solid waste from power plants. The microsphere is filled into plastics to increase anti-friction and anti-pressure performances, the processing flowability, and surface gloss of composites. It makes use of coal ash with high value-added and has great research value.[1] Another example is ultrafine calcium carbonate, which has the advantages of low cost and excellent performances. Ultrafine calcium carbonates with different morphologies have different functions and are applied in various areas. Spherical calcium carbonate is widely used in rubber, paint, and high-end printing ink industries because of its smoothness and good flowability. Nano calcium carbonate products with special morphologies prepared through artificial control of the crystallization of calcium carbonate could have much better application performance and added value.[2]

Inorganic whiskers, a kind of single crystal fibrous or needle-like material that appeared in recent years, are used to reinforce thermosetting resin, thermoplastic resin, rubber, metal, and ceramics through filling and to prepare advanced engineering plastic, composite materials, adhesive, sealant, and paint. The excellent performances of inorganic whiskers have led to their wide application as composite materials.[3,4]

3.1 Whiskers and Inorganic Whiskers

Whiskers are single crystal filamentary materials with definite length-to-diameter (*L/D*) ratio and usually grow under artificially controlled conditions. Whiskers do not have the kind of defects that can occur on macrocrystals because of their tiny diameter. Their atoms are highly ordered and their strength is close to the theoretical value of a perfect crystal. Therefore, whiskers have many outstanding physical, chemical, and mechanical properties, such as light weight, heat resistance, high strength, high modulus of elasticity, high hardness, and so forth that lend excellent physical, chemical, and mechanical performance to plastics, metal, and ceramics reinforced by whiskers.

Hundreds of different kinds of whiskers were developed since they were discovered for the first time in 1948 by scientists from Bell Telephone Company in the United States. Doctor C. Cevans[5] defined whiskers as a kind of fibrous single whiskers with uniform cross section, high-integrity internal structure, *L/D* range above 5–1000, and diameter from 20 nm to 100 μm. However, only whiskers with a diameter of 1–10 μm have particular performances.

There are organic and inorganic whiskers. Organic whiskers include fibrin whiskers, poly(butyl acrylate-phenylethylene) whiskers, and so forth, and are usually used in polymers. Inorganic whiskers include metal whiskers and nonmetal whiskers. Metal whiskers are mainly applied in metal matrix composite material. Ceramic whiskers, a kind of nonmetal whiskers, such as SiC, SiN, MgO, ZnO, $Al_4B_2O_9$, and so forth, possess much better strength and heat resistance performances than metal whiskers. Nonmetal whiskers are widely used in polymer materials and the inorganic whiskers discussed in this book refer mainly to nonmetal whiskers.

3.2 Comparison between Inorganic Whiskers and Previous Inorganic Fillers

The conventional inorganic fillers, such as calcium carbonate (light or ground), mica, garmmite, talcum, and kaolin, are developed quickly in polymers because of their characteristics of lower cost, higher-filling, and better improvement of the mechanic properties, processing performance, and heat property of polymers.

Consider calcium carbonate as an example. The amount of plastic products already exceeds 30 million tonnes a year in China, and that of various calcium carbonate fillers used in plastic industry is more than 2.1 million tonnes per year at present. Along with the rising cost of plastic raw materials (synthetic resin), which increased more than 50% since 2003, plastic manufacturers have paid more attention to inexpensive non-ore powder materials. In particular, calcium carbonate becomes the first choice as an incremental material in plastic processing industries because of its low cost, convenient usage, fewer side effects, and so on.

Although increasing the amount of filling in plastic is better for cost reduction, too much filling will affect the tensile strength and elongation at the break point of products. This is because filling materials are divided in the continuous phase of the matrix resin, and the area of the matrix resin on the stress section is smaller than the material made of pure resin. Under external forces, matrix resin is pulled apart from the surface of filler particles; the total area bearing external forces therefore decreases and the tensile strength of filled plastic will inevitably decrease compared to the unfilled system.

Inorganic whiskers as a single crystal fibrous material have a definite *L/D* ratio and many special properties, such as high strength, high tenacity, heat resistance, wear resistance, corrosion resistance, insulation, conductivity, damping, flame retardance, wave absorption, and so on. The atoms of inorganic whiskers are well arranged when crystallized, which significantly reduces internal defects; therefore their strength is close to the theoretical value of the valence bond between atoms in materials, and whiskers have both reinforcing and toughening effects. Furthermore, inorganic whiskers can be dispersed uniformly in a polymer matrix because of their small size, thus overcome the shortcomings of long fibers, such as uneven dispersion in complex models, poor surface finish, and severe abrasion to models during processing.

Recently, inorganic whiskers with excellent quality and a competitive price, such as calcium sulfate whiskers, calcium carbonate whiskers, and magnesium salt whisker have been developed and applied concomitant with the technical improvement and cost reduction of whiskers. Real polymer/whisker composite material hopefully could be prepared by filling inorganic whiskers to polymers. These new composite materials, called reinforced materials, will combine the rigidity, dimensional stability, and thermal stability of inorganic whiskers with the toughness of polymers, which hopefully will produce new materials that meet the requirements of advanced technology and expand the application fields of existing polymer materials.

3.3 Physical and Chemical Properties of Common Inorganic Whiskers

3.3.1 Silicon Carbide Whiskers

Among existing synthetic whiskers, silicon carbide (SiC) whiskers have the highest hardness, modulus, tensile strength, and heat resistance temperature and are divided into α-SiC and β-SiC according to their different structural characteristics. α-SiC (2H polytype) has a tetragonal structure and scalariform appearance on the surface. β-SiC (3C polytype) has a reed-like morphology and lubricous surface. β-SiC has better mechanical and heat-resisting properties than α-SiC.

Silicon carbide (SiC) whiskers are extremely anisotropic, short fibrous crystals and grow from SiC particles along the [111] plane through catalysis. At present, SiC whiskers are usually prepared by vapor reaction and solid methods, and the latter is more economical and more suitable for industrial production. Figure 3.1 shows the microstructures of SiC whiskers.[6] Table 3.1 shows the chemical and physical properties of SiC whiskers.

Researchers in Japan and the United States have systematically studied the preparation of silicon carbide whiskers. The Institute of Metal Research (IMR), Chinese Academy of Sciences (CAS) synthesized SiC whiskers using silicon oxide (SiO_2) and charcoal under the support of "863" high-technology projects. China Mining University and Shanghai Institute of Ceramics (SIC) of CAS synthesized SiC whiskers by using rice hulls and charcoal + silicon oxide as raw materials, respectively.[7]

As is the case for preparation of SiC, research on SiC-reinforced composites is also focused mainly in Japan and the United States. Some institutes in China, such as Harbin Institute of Technology (HIT), Institute of Metal Research

20 μm

Figure 3.1   SEM image of SiC whiskers.

Table 3.1 Chemical and Physical Properties of SiC Whiskers

SiC Whiskers	Color	Shape	Density (g/cm³)	Diameter (μm)	Length (μm)	Tensile Strength (GPa)	Elasticity Modulus (GPa)	Moh's Scale of Hardness	Melting Point (°C)
α-Type	Ondine	Needle	3.18	0.1–1.0	50–200	12.9–13.7	392	9.2–9.5	1600
β-Type				0.05–1.0	10–200	20.8	480		

(IMR) of Chinese Academy of Sciences (CAS), Shanghai Institute of Ceramics (SIC) of CAS, and Tsinghua University, have also achieved some progress.

3.3.2 Silicon Nitride Whiskers

Silicon nitride (Si_3N_4) whiskers exist in two crystal forms, α-Si_3N_4 and β-Si_3N_4, with the latter used more often in the industry. Si_3N_4 whiskers are normally gray and white in color and have a relative molar mass of 140.28. There are two different morphologies of silicon nitride whiskers: straight and screw-spring forms. Straight whiskers are mainly of the α type and screw-spring whiskers are mostly of the β type. Silicon nitride whiskers have better modulus, strength, hardness, chemical stability, wear resistance, corrosion resistance, high temperature oxidation resistance, and so on than other whiskers and are called the "king of whiskers."

Table 3.2 shows the physical and chemical properties of α and β silicon nitride whiskers produced by Chemical Corporation of Japan and Ube Industries, respectively. Figure 3.2 shows the morphology of silicon nitride whiskers.[8]

3.3.3 Potassium Titanate Whiskers

Potassium titanate whiskers were first explored by the DuPont company in the United States in 1958. Potassium titanate whiskers with different compositions have different performances. The chemical formula of potassium titanate whiskers is $K_2O \cdot nTiO_2$ (n = 1, 2, 4, 6, 8, 10, and 12). The whiskers have a platelike structure when n = 2 and 4, and tunnel structure when n = 6 and 8. The $K_2O \cdot 4TiO_2$ and $K_2O \cdot 6TiO_2$ whiskers have the greatest practical value. Potassium titanate whiskers are generally $K_2O \cdot 6TiO_2$, that is, $K_2Ti_6O_{13}$. Figure 3.3 shows a morphology of $K_2O \cdot 6TiO_2$ whiskers.[9] Table 3.3 shows the chemical and physical properties of $K_2O \cdot 6TiO_2$ whiskers.

Table 3.2 Physical and Chemical Properties of Si$_3$N$_4$ Whiskers

Si$_3$N$_4$ Whiskers	Color	Shape	Density (g/cm^3)	Diameter (μm)	Length (μm)	Tensile Strength (GPa)	Elasticity Modulus (GPa)	Moh's Scale of Hardness	Melting Point (°C)
α-Type	Hoary	Needle	3.18	0.1–1.6	5–200	13.72	382.2	9	1900
β-Type				0.1–5.0	10–50	3.4–10.3			

(a)   (b)

Figure 3.2 SEM images of Si$_3$N$_4$ whiskers. (a) Bar = 5 µm. (b) Bar = 2 µm.

Figure 3.3 SEM image of K$_2$Ti$_6$O$_{13}$ whiskers.

K$_2$O·4TiO$_2$ has good chemical activity. K$_2$O·6TiO$_2$ possesses excellent mechanical and physical properties, stable chemical properties, excellent corrosion resistance, thermal stability, heat insulation, lubricity, high electric insulativity, as well as infrared reflectivity, extremely low thermal conductivity under high temperature and low hardness, electrical resistance 3.3 × 10^{15} Ω·cm, specific heat 0.22 kJ/(kg·K), coefficient of thermal expansion (CTE) 6.8 × 10^{-6}, and ranges of dielectric loss angle tangent 0.06–0.09.[5]

Potassium titanate whiskers can be used as thermal insulation material, electrical insulating material, catalyst carriers, filtration material, friction material, and so on based on their special performances. In addition, they are applied as

Table 3.3 Physical and Chemical Properties of $K_2Ti_6O_{13}$ Whiskers

Color	Shape	Density (g/cm³)	Diameter (μm)	Length (μm)	Tensile Strength (GPa)	Elasticity Modulus (GPa)	Moh's Scale of Hardness	Melting Point (°C)
White or light yellow	Needle	3.3	0.1–1.5	10–100	6.68	274.4	4	1200

conducting materials, ion-exchange materials, and sorbents after being conductively treated with Sb/SnO_2. However, the most attractive and potential application of potassium titanate whiskers is still reinforcement of composites. When potassium titanate whiskers are filled in polymers as reinforcement agents, they not only can significantly improve the abrasion, skid resistance, and dimensional stability of materials, but also have smaller abrasion from equipment and molds because of their low hardness. At present, resin matrixes reinforced by potassium titanate whiskers include polyetheretherketone (PEEK), polyoxymethylene (POM), (poly(butylene terephthalate)) (PBT), polyamide-66 (PA-66), polyamide-6 (PA-6), phenolic aldehyde, special nylon, modified (poly(phenylene oxide)) (PPO), (poly(phenylene sulfide)) (PPS), acrylonitrile butadiene styrene terpolymer (ABS), polyvinyl chloride (PVC), polypropylene (PP), polycarbonate (PC), epoxy resin, and so forth.[7]

3.3.4 Aluminum Borate Whiskers

The chemical formula of aluminum borate whiskers is $xAl_2O_3 \cdot yB_2O_3$. There are three forms of aluminum borate: $9Al_2O_3 \cdot 2B_2O_3$, $2Al_2O_3 \cdot B_2O_3$, and $Al_2O_3 \cdot B_2O_3$. The first two are artificial products while $Al_2O_3 \cdot B_2O_3$ exists in natural minerals. Table 3.4 shows the chemical and physical properties of $9Al_2O_3 \cdot 2B_2O_3$ whiskers, which are chemically resistant to acids or alkalis.[5] Figure 3.4 shows the microstructure of aluminum borate whiskers.[8]

Aluminum borate whiskers are applied widely because of their excellent performances and relatively low cost. Below are their main performance advantages.

1. Higher elasticity modulus than SiC and Si_3N_4 whiskers
2. Higher tensile strength than potassium titanate whiskers
3. Small size: diameter 0.5–1 μm, lengths 10–30 μm
4. Lower hardness than SiC and Si_3N_4 whiskers
5. Good heat resistance, similar to that of potassium titanate whiskers

Table 3.4 Physical and Chemical Properties of Aluminum Borate Whiskers

Color	Shape	Density (g/cm³)	Diameter (μm)	Length (μm)	Tensile Strength (GPa)	Elasticity Modulus (GPa)	Moh's Scale of Hardness	Melting Point (°C)
White	Needle	2.93	0.5–1.0	10–30	7.84	392	7	1200

(a)   (b)

Figure 3.4   (a, b) SEM images of aluminum borate whiskers (bar = 2 μm).

6. High Al_2O_3 content, good covalent with metal, especially aluminum, widely used in aluminum matrix composites
7. Low cost, with a price that is only 1/10–1/30 of that of SiC or Si_3N_4 whiskers
8. Excellent abrasion resistance, good fire resistance, and flame retardance[10]

Based on these excellent performances, aluminum borate whiskers are used not only for thermal insulation material, heat-resistant material, and corrosion-resistant material but also for reinforcing agents in thermoplastic resin, thermosetting resin, ceramics, cement, and metal, and are used mainly to reinforce light metal alloy, ceramic material, resin matrix composite, coatings, and so on.

Reinforcing polymer materials are one of the research hotspots of development of aluminum borate whiskers. Matrix resins that are currently used include PP, polyethylene (PE), PVC, PS, PC, PA, polyimide, and so forth. The polymer composites manufactured with aluminum borate whiskers have excellent performances in strength, rigidity, abrasion resistance, impact resistance, and lubricating property, and are widely used to make automotive brakes, clutch facing, chain wheels, axletrees, and sports equipment. Some researchers from HIT, IMS, SIC, Tsinghua University, Shanghai Jiaotong University (SJTU), and Zhejiang University in China have conducted extensive research and made significant progress in this field.

At present, Chinese researchers have considerable knowledge and experience in the synthesis of aluminum borate whiskers, reinforcing and toughening composites, and industrialization. Before long, aluminum borate whiskers will be used widely in reinforced metal matrixes (aluminum base, magnesium base), ceramic matrix, plastic, glass, fiber, and coatings.[7]

3.3.5 Zinc Oxide Whiskers

Zinc oxide (ZnO) whiskers have two different single crystal structures, tetragonal acicular and hexagonal columnar. The former has an extremely regular shape, every intersection angle of one needle with the other three needles is 109°, and the center is just the center of the tetrahedron. Zinc oxide whiskers are the only whiskers that have a regular three-dimensional structure among currently known inorganic whiskers. Table 3.5 shows the chemical and physical properties of zinc oxide whiskers. Figure 3.5 shows the micromorphology of zinc oxide whiskers.[8]

Zinc oxide whiskers possess the following attractive physical properties[1]:

1. Highly regular morphology and size. General zinc oxide whiskers are needle-like or particles with different sizes, while polygonal zinc oxide whiskers have a uniform structure and size.
2. Good isotropy. The composites made with zinc oxide whiskers have good isotropic mechanical, electrical, and optical properties because of the isotropic structure and properties of zinc oxide whiskers.
3. Semiconductor performance. The n-type semiconductor (resistivity = 7.14 $\Omega \cdot cm$) is formed when polygonal needles of polygonal zinc oxide whiskers overlap with each other. The resin matrix composites reinforced by polygonal zinc oxide whiskers could be used to make electronic

Table 3.5 Physical and Chemical Properties of Zinc Oxide Whiskers

Color	Shape	Density (g/cm³)	Diameter (µm)	Length (µm)	Tensile Strength (GPa)	Elasticity Modulus (GPa)	Moh's Scale of Hardness	Melting Point (°C)
White	Tetragonal acicular	5.78	0.3–2.0	2–300	9.8	350	4	1720

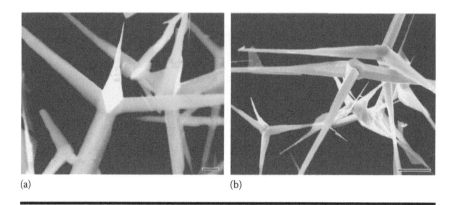

(a)   (b)

**Figure 3.5   SEM images of zinc oxide whiskers. (a) Bar = 1 μm.
(b) Bar = 5 μm.**

    components requiring electric conduction or an electro-
magnetic shield.
4. Ultraviolet absorption. Zinc oxide whiskers have obvious
   ultraviolet absorption properties, which could be used to
   improve the aging resistance of polymers.
5. High density. Zinc oxide whiskers can be used to make
   composites with sound insulation, shock absorption, and
   vibration resistance properties because of their high relative density (5.78 g/cm³).

    Like other acicular whiskers, general acicular zinc oxide
whiskers are mainly used as reinforcing agents in composites;
tetragonal acicular ZnO whiskers are also used to prepare
functional composites because of their unique structure. They
are employed to improve some properties of composites, such
as tensile strength, flexure resistance, wear resistance at room
temperature, as well as chemical and dimensional stability at
high temperature.

    Because zinc oxide whiskers have small volume electri-
cal resistivity (<50 Ω/cm), high real density (5.8 g/cm³), and
low packing density (0.01–0.5 g/cm³), they could give composite materials sound absorption, shock absorption, and
vibration resistance properties, and can be used in audio

machines. Zinc oxide whiskers also have antibacterial and ultraviolet and infrared light absorption properties, and it has been proved that zinc oxide whiskers/resin matrix composites could kill 99% *Escherichia coli*, *Staphylococcus aureus*, and *Pseudomonas aeruginosa*. In addition, zinc oxide whiskers could also be used to make antibacterial refrigerators, telephones, grocery bags, floorboards, and so on. Furthermore, zinc oxide whiskers can be applied in audio machines, sterilization materials, and analytical equipment (e.g., atomic force microscope [AFM] and scanning tunneling microscope [STM]).[7]

3.3.6 Magnesium Oxide Whiskers

The molecular weight of magnesium oxide whiskers (MgO) is 40.31. The micromorphology is fibrous or whiskers; the cross-sectional shape varies with raw material and synthetic process, and includes mainly round, cross, and H shapes. They have good heat resistance, electrical insulativity, heat conduction (three times that of alumina oxide), heat stability, and reinforcement and toughness properties. Table 3.6 shows the chemical and physical properties of magnesium oxide whiskers. Figure 3.6 shows the micromorphology of magnesium oxide whiskers.[8]

3.3.7 Calcium Carbonate Whiskers

Calcium carbonate ($CaCO_3$) whiskers are of several different types, such as calcite, aragonite, and vaterite. Aragonite whiskers are a new acicular material synthesized in recent years. They are a white powder and appear as acicular single crystals under a microscope.

Calcite type calcium carbonate whiskers have a single crystal crystalline form and have almost no internal defects; therefore they possess some special characteristics, such as high intensity, high modulus, good heat resistance and heat

Table 3.6 Physical and Chemical Properties of Magnesium Oxide Whiskers

Color	Shape	Density (g/cm³)	Diameter (μm)	Length (μm)	Tensile Strength (GPa)	Elasticity Modulus (GPa)	Moh's Scale of Hardness
White	Fibrous	3.58	0.5–5.0	200–2000	980	310.1	6

(a)　　　　　　　　　　　(b)

Figure 3.6　(a, b) SEM images of magnesium oxide whiskers (bar = 100 μm).

insulation, and so forth. Their advantages include low cost, high whiteness, and large filling content, and hopefully can be widely used in plastic commodities due to their biodegradability.[11,12] Table 3.7 shows the chemical and physical properties of calcium carbonate whiskers. Table 3.8 shows the enterprise standard of calcium carbonate whiskers produced by Qinghai Haixing Technology Development Co., Ltd. (QHTD).[5] Figure 3.7 shows the micromorphology of calcium carbonate whiskers.[8]

Table 3.7　Physical and Chemical Properties of Calcium Carbonate Whiskers

Color	Shape	Density (g/cm³)	Diameter (μm)	Length (μm)
White	Fibrous	2.86	0.5–1.0	20–30

Table 3.8　Enterprise Standard (Q/HXK002-2003) of Calcium Carbonate Whiskers

Items	Indexes
$CaCO_3$ content (dry basis, %)	96 ± 1.5
Lengths of whiskers (μm)	10–40
Diameters of whiskers (μm)	≤2
Moisture	≤0.5

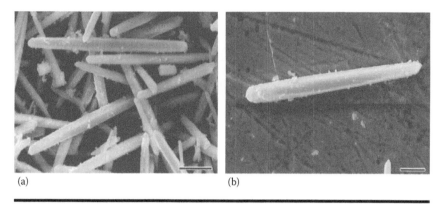

(a)   (b)

Figure 3.7   SEM images of calcium carbonate whiskers. (a) Bar = 10 μm. (b) Bar = 5 μm.

3.3.8 Calcium Sulfate Whiskers

Calcium sulfate whiskers, also known as gypsum whiskers, are fibrous single crystals of anhydrous calcium sulfate and are classified into anhydrite ($CaSO_4$), hemihydrate ($CaSO_4 \cdot 0.5H_2O$), and dihydrate ($CaSO_4 \cdot 2H_2O$) calcium sulfate whiskers.

Anhydrite calcium sulfate whiskers ($CaSO_4$; molecular weight, 136.14) are white fluffy solids in appearance and fibrous or acicular monocrystals under a microscope. They are divided into medium fibers and thin fibers. The diameters of medium fibers are 1–2 μm, lengths are 30–150 μm, and L/D ratios are 20–100. The diameters of thin fibers are 0.1–1.5 μm, lengths are 20–120 μm, and L/D ratios are 30–200. They can be used as reinforcing agents with medium intensity because of the following properties: relative density is 2.96, Moh's hardness is 3, melting point is 1450°C, heat stability point is 1000°C, tensile strength is 2.058 GPa, and elasticity modulus is 176.4 GPa. The reinforcing effect of thin fibers is similar to that of other high-energy fibers.

Hemihydrate calcium sulfate whiskers ($CaSO_4 \cdot 0.5H_2O$; molecular weight, 145.15) are white fluffy solids in appearance and fibrous or acicular monocrystals under a microscope, with diameter, length, hardness, heat resistance, and intensity all between those of anhydrite and dihydrate calcium sulfate whiskers. Hemihydrate calcium sulfate whiskers will dehydrate to amorphous calcium sulfate powder at 160°C and therefore cannot be used as reinforcing and toughening agents above 160°C.

Dihydrate calcium sulfate whiskers ($CaSO_4 \cdot 2H_2O$; molecular weight, 172.18) are white fluffy solids in appearance and fibrous or acicular crystals long fibers under a microscope. Their diameters are 10–50 µm, lengths are >500 µm, and *L/D* ratios are 20–100. They have relatively poor hardness, heat resistance, and intensity. These whiskers will dehydrate at room temperature and transform to amorphous calcium sulfate particles at about 110°C. Thus, applications of dehydrate calcium sulfate whiskers in composites are limited.[5]

In general, anhydrite calcium sulfate whiskers are more competitive than other whiskers because of their good properties, such as good chemical stability, good heat resistance, chemical corrosion resistance, high intensity, good friction resistance, easy surface treatment, strong compatibility with polymers, less toxicity, low cost (only 1/200–1/300 that of SiC whiskers), and so on. Anhydrite calcium sulfate whiskers can replace asbestos as reinforcing agents.[6,13] Figure 3.8 shows the microstructure of anhydrite calcium sulfate whiskers.[14] Table 3.9 shows the enterprise standard of calcium sulfate whiskers produced by Qinghai Haixing Technology Development Co., Ltd.[5]

30 µm

Figure 3.8   SEM image of calcium sulfate whisker.

Table 3.9   Enterprise Standard of Calcium Sulfate Whiskers

Items	Stubby Whiskers	Slender Whiskers
Appearance	White fluffy solid, needle-like under a microscope	White fluffy solid, needle-like under a microscope
Relative molecular weight	136.14	136.14
Diameters (μm)	1–2	0.1–1.5
Lengths (μm)	30–50	20–120
L/D ratio	20–100	30–200

3.3.9 Magnesium Borate Whiskers

Magnesium borate ($Mg_2B_2O_5$) whiskers, also known as magnesium diborate whiskers, were discovered for the first time in a natural suanite mine in South Korea in 1953. Sheet and columnar magnesium borate crystals were synthesized during the 1960s. These whiskers are deemed to be the most hopeful and applicable whiskers used in composite materials today because of their low cost (1/20–1/30 that of SiC whiskers) and prominent mechanical performances. Table 3.10 shows the chemical and physical properties of magnesium borate whiskers. Figure 3.9 shows the microstructure of these whiskers.[15]

Magnesium borate whiskers have good mechanical performances and their reinforced plastic composites have good fluidity, lubricous surface, good moldability, and dimensional stability, and they therefore can be used to manufacture small parts and ultrathin parts, such as parts of watches and cameras. Moreover, magnesium borate whiskers can be used in some anti-wearing and heat-resisting parts, such as engine pistons, connecting rods, cylinders, automotive brakes, chain wheels, axletrees, and some sports goods because of

Table 3.10 Physical and Chemical Properties of Magnesium Borate Whiskers

Color	Shape	Density (g/cm³)	Diameter (μm)	Length (μm)	Tensile Strength (GPa)	Elasticity Modulus (GPa)	Moh's Scale of Hardness	Melting Point (°C)
White	Needle	2.91	0.2–2.0	10–50	3.92	264.6	5.5	1360

Figure 3.9   SEM image of magnesium borate whiskers.

lightweight, high toughness, anti-wearing properties, and chemical stability.

The low cost of raw materials, simple manufacturing technique, low facility demand, and easy operation process make the cost of magnesium borate whiskers much lower. Thus, magnesium borate whiskers are a kind of new cost-effective reinforcing agent with potentially wide applications after aluminum borate whiskers, and can be used in aluminum, magnesium, and their alloys and in engineering plastics.[1,3,16]

3.3.10 Magnesium Sulfate Whiskers

Magnesium sulfate whiskers ($MgSO_4 \cdot 5Mg(OH)_2 \cdot 3H_2O$) are referred to as magnesium hydroxide sulfate hydrate whiskers. Similar to other whiskers, they are tiny white needle-like single crystal fibers with diameters ranging from 0.5 to 1.0 μm, lengths ranging from 20 to 80 μm, relative density of 2.3 g/cm³, and heat-resisting temperature of 300°C. The whiskers have excellent physical performances, such as high intensity, high modulus, and good electric properties.

Magnesium sulfate whiskers have expansion coefficients similar to those of plastics, and good compatibility with plastics, and their composite materials have good physical and mechanical properties. Furthermore, they have properties such as nontoxicity, low smoke generation, and fire resistance.

Figure 3.10   SEM images of magnesium sulfate whiskers. (a) ×1000. (b) ×5000.

Magnesium sulfate whiskers have both reinforcing and flame retardant effects when filled into plastic or resin matrix to make composite materials. Composites manufactured with magnesium sulfate whiskers can be used in some parts of trains, trucks, and other motorcars because of their excellent chemical and physical performances.[17] Figure 3.10 shows the microstructure of magnesium sulfate whiskers.[18]

3.3.11 Mullite Whiskers

Mullite whiskers ($3Al_2O_3 \cdot 2SiO_2$) have the following characteristics: diameters range from 0.5 to 1.0 µm, lengths range from 7.5 to 20 µm, density is 2.91 g/cm³, melting point is >2000°C, and heat-resistant temperatures range from 1500 to 1700°C.

Mullite is a catenarian silicate mineral ($Al_xSi_{2-x}O_{5.5-0.5x}$) belonging to the rhombic system and is an excellent refractory material. The biggest advantage of mullite as a metal reinforced material is its chemical inertness. Mullite whiskers can be used as ultra high-temperature thermal insulation material and are used in industrial high-temperature furnaces and reinforcing material of fiber-reinforced metal. The industrial production of mullite whiskers is concentrated mainly in Japan

Table 3.11   Chemical Compositions of Mullite Whiskers

	Manufacturers	Al_2O_3 (%)	MgO (%)	SiO_2 (%)	B_2O_3 (%)
Short whisker	Imperial Chemical Industries (ICI) of British	95	5	–	–
	Denki Kagaku	80	20	–	–
Long whisker	Mitsubishi Chemicals Corporation	62	14	24	–
	DuPont USA	99	–	–	1
	Sumitomo Chemical	85	15	–	–
	Denki Kagaku	80	20	–	–

and the components of mullite whiskers from different manufacturers vary. Table 3.11 shows the chemical compositions of mullite whiskers from different manufacturers.[5]

3.3.12 Other Inorganic Whiskers

In addition to the aforementioned whiskers, there are many others whiskers, such as graphite (C) whiskers, barium titanate ($BaTiO_3$) whiskers, titanium boride (TiB_2) whiskers, titanium oxide (TiO_2) whiskers, SiO_2-MgO-CaO (SMC) whiskers, and so forth. Table 3.12 shows the chemical and physical properties of different whiskers.

The average diameter, length, and *L/D* ratio of barium titanate whiskers produced by Otsuks, Co. Ltd. of Japan are 0.3 μm, 3 μm, and about 10, respectively. These whiskers can be used in piezoelectric composites because of their piezoelectricity when the crystal size is greater than 0.27 μm.[19]

The SMC whiskers composed of SiO_2, MgO, and CaO have a light yellow appearance and needle-like or rod morphology with diameters ranging from 0.5 to 2 μm and an *L/D* ratio above 16.[20] Figure 3.11 shows the microstructure of SMC whiskers.[21]

Table 3.12 Physical and Chemical Properties of Other Inorganic Whiskers

Whisker Types	Color	Shape	Density (g/cm³)	Diameter (μm)	Length (μm)
Graphite	–	Needle or fibrous	2.25	0.3–1.0	5–100
Barium titanate	White	Needle	5.5	0.2–0.5	10–20
Titanium boride	–	Needle	4.48	0.2–0.8	Hundreds of micrometers
Titanium oxide	–	Needle	4.2	0.5–0.1 0.15–0.05	3–6 4–12
SMC	Yellow	Needle or rod	–	0.5–2.0	8–32

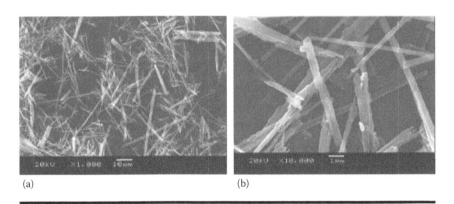

(a)　　　　　　　　(b)

Figure 3.11 SEM images of SMC whiskers. (a) ×1000. (b) ×1000.

3.4 Surface Modification Methods of Inorganic Whiskers

The ideal conditions of whisker-reinforced resin are as follows: whiskers are dispersed evenly in a matrix, separated from each other, and are well wetted by matrix resin. Surface modification of whiskers is needed to improve the dispersion of inorganic whiskers in an organic matrix and the interaction between whiskers and matrix.

The surface modification methods of inorganic whiskers mainly include the wet method, dry method, *in situ* polymerization, sol–gel method, surface coating and coupling agent modification, self-assembly method, and so on.

3.4.1 Wet Method

In the wet method, surface treatment is conducted in the emulsion or aqueous solution of the treating agent, and the filling surface and molecules of the treating agent are bonded together by a chemical effect and surface absorption of the filling material. This method is always used to treat coupling agents that can dissolve in water or form an emulsion, such as stearate, silane coupling agent, and titanate coupling agent.

The methods are different when different inorganic whiskers are treated with different coupling agents because of not only the characteristics of the coupling agents but also the surface property of the inorganic whiskers and the bond formation of both. For example, high temperature and hydroxyl groups absorbed on the whisker surface are helpful for esterification when inorganic whiskers are treated with stearate and its salts. Silane coupling agents are always prepared as dilute solutions using water, alcohol (ethanol, isopropanol), or a mixture of both to treat inorganic whiskers. When preparing solutions of silane coupling agents, except for ammonia hydroxyl silane, acetic acid is always added as a hydrolysis catalyst, and the pH is adjusted to 3.5–5.5. Long-chain alkyl silane and phenyl silane are unsuitable for preparation into aqueous solution because of their poor chemical stability. Chlorosilane and ethoxysilane are also not suitable for preparation into aqueous solution or a water and alcohol mixture solution owing to their severe condensation reaction during a hydrolysis reaction. A silane coupling agent with poor water solubility should be prepared as an emulsion after the addition of nonionic surfactant.

Han et al.[22] modified basic magnesium sulfate whiskers using sodium stearate. The method is as follows. First, measure 50 g of basic magnesium sulfate whiskers and prepare a 5% (mass fraction) mixture of whiskers and water. Second, blend the mixture evenly and then keep it at constant temperature at 80°C in a water bath. Third, add sodium stearate to the mixture and stir for 30 minutes. Products are obtained after filtration and low-temperature drying. Finally, strong chemical adsorption between whiskers and sodium stearate develops after modification.

Li et al.[23] modified potassium titanate whiskers using a KH-550 silane coupling agent. The process is as follows. First, prepare a solution with KH-550 (1 wt% whisker concentration) and ethanol. Second, add the whiskers to the solution and stir for 30 minutes at high speed. Products are obtained after filtration and drying at 120°C for 1 hour. Moreover, Zhang et al.[24] modified potassium titanate whiskers using the silane coupling agent Z-6030. The method is as follows. First, measure ethanol and distilled water (mass ratio of 9:1) and prepare into a mixture solution. Second, add glacial acetic acid to adjust the pH to 4.0–5.0. Third, add the coupling agent Z-6039 (5 wt% of the mixture) into the mixture solution and stir for an hour to prehydrolyze Z-6039. Fourth, the whiskers are dispersed evenly in distilled water (5 times the whisker mass) by stirring and ultrasonic surge for 5 minutes. Then, drip the prepared coupling agent mixture into the whisker slurry and stir for 10 minutes and allow to stand for 30 minutes. The final products are obtained after filtration and drying at 120°C for 24 hours.

Liu et al.[25] treated zinc oxide whiskers with different kinds of silane coupling agents and titanate coupling agents. The process using silane coupling agent is as follows. Add silane coupling agent to hydrochloric solution of water–ethanol or water–acetone and hydrolyze for 20 to 45 minutes at a pH of 3–5. Then add zinc oxide whiskers into the solution and stir for 30 to 60 minutes in a water bath at a constant temperature. The final products are prepared after filtration, washing,

Table 3.13 Best Treatment Parameters of Different Silane Coupling Agents

Coupling Agent	pH	Solvent	Hydrolysis Temperature (°C)	Dispersion Time (min)
KH-570	3.5	Water–acetone	60	50
A151	4.0	Water–ethanol	70	30
ND-42	5.0	Water–actone	80	45

drying at 80°C, and activating at 150°C for 8 hours. The process using titanate coupling agent is as follows. Add titanate coupling agent to acetone and then add whiskers after the coupling agent is completely dissolved. Then, the mixture is dispersed for 45 minutes, washed, and dried. Table 3.13 shows the best treatment parameters of three kinds of silane coupling agents when surface modifying zinc oxide whiskers.

3.4.2 Dry Method

In the dry method, treating agents are coated evenly on the surface of dry filling powder under high-speed stirring and mixing in a high-speed mixer. Aluminate and titanate coupling agents are often treated using the dry method. Sometimes, industrial white oil or liquid paraffin is added to dilute the surface treatment agent because of a very small concentration of the treatment agent (0.15%–3%).

The treatment process of calcium carbonate whiskers using the dry method is as follows.[26] Measured materials are mixed in a high-speed mixer at slow speed for 1 minute and at high speed for 1 minute. Wang et al.[27] added calcium sulfate whiskers into a high-speed mixer, then heated and stirred. They then added surface treatment agent diluted with liquid paraffin at 100°C, stirred for 6 minutes, and obtained modified calcium sulfate whiskers.

3.4.3 In Situ *Polymerization Method*

In the *in situ* polymerization method, first inorganic whiskers are evenly dispersed in monomers and then polymerized using a method similar to bulk polymerization. In this method, surface modification of whiskers and preparation of composites take place at the same time.

Chen et al.[28] treated zinc oxide whiskers via *in situ* polymerization. The first step is the dispersion process of zinc oxide whiskers. First expose zinc oxide whiskers to air for 24 hours, after which a large number of zinc oxide groups are produced. Then dry for 2 hours at 120°C. Then submerge the whiskers into an ethanol solution of APTS (γ-aminopropyltriethoxysilane) and stir for 10 hours at 80°C. The sample is treated in an ultrasonic tank and then dried. The second step is the polymerization process between the treated whiskers and matrix. First, a certain amount of zinc oxide whiskers, aniline, and ethanol are mixed and stirred in a magnetic stirrer. Then 0.01 mol/L hydrochloric acid is added dropwise while stirring to trigger the polymerization reaction, which is done in an ice bath at 0°C–2°C for 10 hours. Then the polymerized products are treated for 10 minutes in a centrifuge at 3000 rpm and dried for 24 hours in a vacuum furnace.

Wan et al.[29] directly added whiskers into absolute ethyl alcohol solution containing polyethylene glycol. The ratio of whiskers, polyethylene glycol, and absolute ethyl alcohol is 10:3:300. Stir for 5 hours. Then the dispersed whiskers are filtered and dried, and modified whiskers are obtained. Then *in situ* polymerization between the prepared whiskers and matrix will occur during the preparation process of composites.

3.4.4 *Surface Coating and Coupling Treatment*

In surface coating and coupling treatment, first the whisker surface is coated with a layer of inorganic substance, such as silicon oxide film or aluminum oxide film, and then whiskers are surface modified using coupling agents.

This method was first employed in fusing nano-SiO_2 on the surface of silicon-containing whiskers, such as β-Si_3N_4 and SiC whiskers, to form silica-fused ceramic whiskers, which are used in composite resins for tooth repair. The distribution and fusion of nano-SiO_2 on the surface of silicon-containing whiskers increase the silicification property of the whisker surface, which not only is better for the dispersion of whiskers in the resin matrix but also increases the combination between the rough surfaces of whiskers and the resin matrix, thus significantly improving the mechanical properties of composite materials.[30]

The coating process of whiskers is as follows.[31]

1. *Coating silicon oxide film.* Whiskers (10 wt%) are mixed with water and quickly dispersed. Sodium silicate, measured according to 5%–10% of coating content, is dissolved in water and added slowly into the whisker slurry with stirring. Then the pH is adjusted to 9.5 by slowly adding diluted sulfuric acid and the mixture is kept at insulation aging for 5 hours. Coated whiskers are obtained after filtration, washing, and drying.

2. *Coating aluminum oxide film.* The process and concentration are the same as above. Whiskers (10 wt%) are mixed with water and quickly dispersed. Sodium aluminate, measured according to 5%–10% of coating content, is dissolved in water and added slowly into the whisker slurry with stirring. Then the pH is adjusted to 5.0 by slowly adding diluted sulfuric acid and the mixture is kept at insulation aging for 5 hours. Coated whiskers are obtained after filtrating, washing, and drying.

Furthermore, coating silicon oxide film also can be achieved by using the TEOS (tetraethylortho) sol–gel method. Aluminum borate (AlBw) whiskers are added to absolute ethyl alcohol and dispersed in an ultrasonic cleaner and then moved into a rotary evaporator. Then, TEOS and the mixture of water and ammonia are added to the whisker slurry according to a certain ratio. The amount of TEOS is determined by the weight ratio of SiO_2 hydrolyzed from TEOS to AlBw. The mixture is stirred and heated to 70°C and stirring is continued for 6 hours. After the reaction, nitrogen is immediately let in and the mixture is evaporated until the solvent is completely evaporated in a water bath at 78°C–80°C. The final products are obtained after sintering for 2 hours at 800°C with a heating rate of 250°C/h in a resistance furnace and cooled in the furnace.[30]

3.4.5 Self-Assembly Method

The self-assembly method has attracted much attention because it can form monomolecular films on the surface of many matrixes, such as silicon oxide and titanium oxide. Considering potassium titanate whiskers as an example, the similar surface structure (Ti–O) between potassium titanate whiskers and titanium oxide means there is some comparability between them.

Wang et al.[32] surface modified potassium titanate whiskers (PTW) using the self-assembly of octadecyl trichloro silane (OTS). The process is as follows. Add 0.5 mL of OTS to the suspension liquid of whiskers (10 g) and heptane (200 mL) with stirring and continue stirring for 20 minutes after the whiskers are evenly dispersed in the solution. Then the mixture is filtered and washed several times to remove dissociative OTS and dried in a vacuum.

After surface modification, whiskers have much better lipophilic and hydrophobic properties, and the alkyl chains of OTS absorbed on the whisker surface are inclined while OTS usually is distributed unequally on the matrix surface. The self-assembly mechanism of OTS is as follows. First, Si–Cl bonds are changed to Si–OH bonds when OTS and trace water on the whisker surface are hydrolyzed. Then, Si–OH

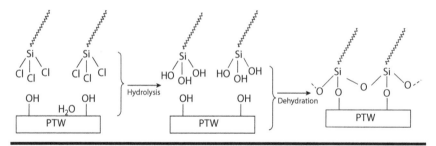

Figure 3.12 Self-assembling process of OTS on the surface of PTW.

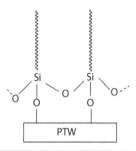

Figure 3.13 OTS is perpendicular to the surface of PTW.

bonds and OH bonds of the whisker surface are dehydrated. Finally, alkyl chains of OTS and the whisker surface form covalent bonds, which means accomplishment of the surface modification. Figure 3.12 shows the self-assembly process.

If modified whiskers are dissolved in organic solvent, the alkyl chains of OTS will spread into an organic solvent and absorb vertically on the surface of whiskers, as shown in Figure 3.13.

3.5 Surface Modification Evaluation of Inorganic Whiskers

Surface modification evaluation of inorganic whiskers can be achieved by measuring the activation index, adsorbing capacity, surface free energy, surface wettability, and so forth of modified whiskers, or by researching the interaction and micro-morphology of the modifying agent and whisker surface

using modern analyzing techniques such as infrared spectroscopy and so on.

There are two different methods of surface modification
evaluation of inorganic whiskers: the direct method and the
indirect method.

3.5.1 Direct Method

The direct method is used to evaluate surface modification
results by measuring surface property changes of whiskers
before and after modification.

Inorganic whiskers will sink naturally in water because of
their relative high density and surface polarity. Surface modifiers are usually water-insoluble organic surfactants. Modified
whiskers will float in water just like oil slicks, because of their
strong hydrophobicity after being treated with surfactants. This
phenomenon could reflect a surface modification effect of
inorganic whiskers. Surface modification results can be evaluated by detecting the activation index, oil adsorbing capacity,
surface wettability, and so forth.

3.5.1.1 Activation Index

In actual research, the activation index (H, mass ratio of the
floating portion to total samples) is used to evaluate the surface modification results of whiskers (refer to the national standard GB/T 19281-2006). The detection process is as follows.[33]
Add a certain amount of modified whiskers into 100 mL of
deionized water and let it sit until clear after stirring for 1–2
minutes. Then remove the floating portion on the solution
surface and collect the sunken powders at the bottom. The latter are filtered, dried, and weighed and the activation index is
calculated using the following equation:

$$H = \left(m - \frac{m_1}{m} \right) \times 100\% \qquad (3.1)$$

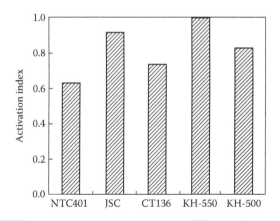

Figure 3.14 **Activation indexes of CaSO₄ whiskers modified by different coupling agents.**

In this formula, H is the activation index (%); m is the total sample (g); m_1 is the sample sunken at the bottom (g).

Zou et al.[34] modified calcium sulfate whiskers using three kinds of titanate coupling agents (NTC401, CT136, and JSC) and two kinds of silane coupling agents (KH-550 and KH-560) with a concentration of 8 wt%, respectively. The activation indexes of modified whiskers are shown in Figure 3.14.

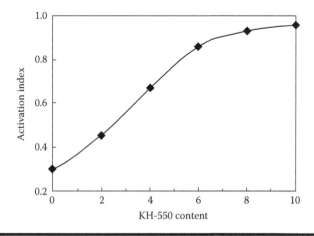

Figure 3.15 **Activation indexes of CaSO₄ whiskers affected by KH-550 concentration.**

As is shown in Figure 3.14, KH-550 has the best modification effect on calcium sulfate whiskers with an activation index (99.6%) higher than that of other coupling agents. The sample modified with KH-550 cannot be wetted by water. Figure 3.15 shows the modification results of calcium sulfate whiskers affected by KH-550 concentration. Modification results become much better with an increase of KH-550 concentration. But the rising trend of activation index is no longer obvious when the concentration of KH-550 is more than 8 wt%.

3.5.1.2 Measurement of Oil Adsorbing Capacity (Refer to GB/T 19281-2003 Standard)

Dioctyl phthalate (DOP) filled in a drop bottle is dropped into 5 g of a modified whisker sample (precise to 0.01 g), which is placed on a glass or ceramic board. The sample needs to be milled using palette knives while dripping. With the dripping of DOP, the appearance of the mixture changes from a dispersed powder to a reunited group. Stop dripping until all samples are wetted by DOP and the mixture of whiskers and DOP becomes a group. The whole process should be finished within 90 minutes.

Oil adsorbing capacity is marked as ω_η and denoted as DOP weight (g) adsorbed by 100 g of activated calcium carbonate. The oil adsorbing capacity is calculated as follows:

$$\omega_\eta = \left[\frac{m_1 - m_2}{m} \right] \times 100 \tag{3.2}$$

In this formula, m_1 is the total weight of DOP and the drop bottle before dripping DOP; m_2 is the total weight of DOP and the drop bottle after dripping DOP; m is the total weight of modified whiskers.

Take the average value of several parallel determinations as the final result. Furthermore, the absolute difference of parallel determinations should not be more than 1.0 g/100 g.

3.5.1.3 Contact Angle, Surface Free Energy, and Polar Component

The contact angle (θ), another criterion for hydrophilicity and lipophilicity, is employed to evaluate the wettability of liquid to solid. Generally, the solid surface is considered to be wetted by liquid when $\theta < 90°$; the solid surface is not considered to be wetted by liquid when $\theta > 90°$; and the solid surface is wetted completely by liquid when $\theta = 0°$. As for inorganic whiskers, the bigger the contact angle between distilled water and whiskers, the better is the hydrophobicity of whiskers.

The contact angle is measured mainly by using a contact angle meter. The measuring process is as follows. Keep 3 g of modified whiskers in a tablet machine at a certain pressure and for a certain time until the whiskers are pressed as a round wafer with a diameter of 12 mm. Then measure the contact angles of a water drop using a contact angle meter. The final contact angle of the sample is the average value of three results.

Liu et al.[25] treated zinc oxide whiskers using several different coupling agents. Table 3.14 shows the contact angles using different coupling agents with a concentration of 4 wt%.

As is shown in Table 3.14, contact angles of modified whiskers can be affected by coupling agent types, but all exhibit hydrophobicity. Figure 3.16 shows the relationship between contact angles and coupling agent concentrations when whiskers are treated with the silane coupling agent KH-570. The contact angles increase with the increasing concentration of the coupling agent. Whiskers exhibit hydrophobicity when the

Table 3.14 Contact Angles Using Different Coupling Agents with a Mass Fraction of 4 wt%

Coupling Agents	KH-550	A151	ND-42	NDZ-130
θ (°)	120.0	129.4	119.2	106.5

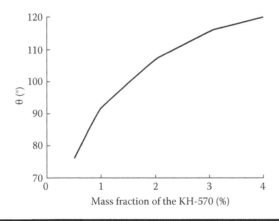

Figure 3.16 **Relationship between KH-570 concentrations and contact angles.**

coupling agent concentration is 1%. However, contact angles change much less when the concentration is higher than 3%.

Wang[32] modified potassium titanate whiskers (PTW) using octadecyl trichloro silane (OTS) and calculated the surface free energy and polar component of modified whiskers by determining contact angles (Table 3.15) of benzene, glycol, and water on modified whisker wafers and using the van Oss–Chaudhury–Good (OCG) equation.

OCG equation:

$$(1 + \cos \theta)\gamma_L = 2\left(\sqrt{\gamma_S^{LW} \gamma_L^{LW}} + \sqrt{\gamma_S^+ \gamma_L^-} + \sqrt{\gamma_S^- \gamma_L^+}\right) \quad (3.3)$$

Table 3.15 **Surface Energy Components of Benzene, Ethanol, and Water and Their Contact Angles on an OTS/PTW Wafer**

Liquid	γ_L (mJ/m^2)	γ_L^{LW} (mJ/m^2)	γ_L^{AB} (mJ/m^2)	γ_L^+ (mJ/m^2)	γ_L^- (mJ/m^2)	θ
Benzene	28.9	28.9	0	0	0	13.6°
Ethanol	48.0	29.0	19.0	1.92	47.0	67°
Water	72.8	21.8	51.0	25.5	25.5	97°

$$\gamma_S = \gamma_S^{LW} + \gamma_S^{AB} = \gamma_S^{LW} + 2\sqrt{\gamma_S^+ \, \gamma_S^-} \qquad (3.4)$$

where

θ = the contact angle

γ_L = the surface energy of the liquid

γ_S = the surface energy of the solid

γ_S^{LW} = the Lifshitz-Van der Waals component of solid surface energy, also called the dispersive component

γ_L^{LW} = the dispersive component of liquid

γ_L^{AB} = the polar component of liquid

γ_S^{AB} = the Lewis acid–base component of solid surface energy, also called the polar component

γ_S^+ = the Lewis electron-acceptor component of solid

γ_L^+ = the Lewis electron-acceptor component of liquid

γ_S^- = the Lewis electron-donor component of solid

γ_L^- = the Lewis electron-donor component of liquid

The calculation results are shown in Table 3.16.

As shown in Table 3.16, the whiskers modified with OTS have lower surface energy, lower polar component, and better hydrophobic and lipophilic properties than untreated whiskers. The dispersive component ratio of potassium titanate whiskers modified with OTS to OTS molecules is much closer to the dispersion value of the $-CH_2-$ group rather than the $-CH_3-$ group, which indicates that the whisker surface modified with

Table 3.16 Surface Energy, Dispersive, and Polar Components of Unmodified/Modified Potassium Titanate Whiskers

Sample	γ_S (mJ/m²)	γ_S^{LW} (mJ/m²)	γ_S^{AB} (mJ/m²)
Unmodified whiskers	69.2	20.1	49.1
Modified whiskers	29.0	28.1	0.9

OTS has mainly –CH$_2$– groups and inclined alkyl chains of OTS. Furthermore, the polar component of whiskers modified with OTS decreases almost to zero, indicating that the surface of modified whiskers has stronger hydrophobicity and exhibits complete lipophilicity.

In addition, the adsorption work and interfacial tension can be calculated as follows:

$$W_{ad} = (1 + \cos \theta)\gamma_L \tag{3.5}$$

$$\gamma_{SL} = \gamma_S + \gamma_L - 2\left(\sqrt{\gamma_S^{LW}\, \gamma_L^{LW}} + \sqrt{\gamma_S^{AB}\, \gamma_L^{AB}}\right) \tag{3.6}$$

where

W_{ad} = the adsorption work of solid to liquid

γ_{SL} = the interfacial tension between solid and liquid

Table 3.17 shows the adsorption work and interfacial tension of potassium titanate whiskers before and after modification with OTS with water and benzene. After modification with OTS, the adsorption work of modified whiskers to oil is lower than to water, and the interfacial tension of whiskers to water increases while that to oil is close to zero. These prove that whiskers treated with OTS have more typical hydrophobicity and lipophilicity.

Table 3.17 Adsorptive Work and Interfacial Tension of Potassium Titanate Whiskers before and after Modification with Water and Benzene

Sample	W_{ad} (Water/Solid) (mJ/m²)	W_{ad} (Benzene/Solid) (mJ/m²)	γ_{SL} (Water/Solid) (mJ/m²)	γ_{SL} (Benzene/Solid) (mJ/m²)
Unmodified whiskers	−145.6	41.9	0.0525	50.0
Modified whiskers	63.9	57.0	38.8	≈0

3.5.1.4 Infrared Spectroscopic Analysis

The infrared spectrum, also known as molecule vibration and rotation spectrum, is used in fundamental research of molecular structure and analysis of chemical components, and the latter is the widest application of the infrared spectrum. The structure of an unknown sample can be deduced according to the position and shape of absorption peaks in the spectrum and contents of the components of the mixture can be detected by the strength of the characteristic peaks. Infrared spectrometry has become the most widely used analysis and test instrument because of its analytical characteristics of high efficiency, high sensitivity, lower sample quantity, and good sample applicability.

The infrared spectrum can also be used to evaluate the modification effects of inorganic whiskers by an organic modifier. As we can see in Figure 3.17, the infrared spectrum of zinc oxide whiskers (ZnOw) shows remarkable changes before and after being modified with different kinds of titanate coupling agents and silane coupling agents. After zinc oxide whiskers are modified with the titanate coupling agent NDZ-130, new infrared adsorption peaks at 1396 cm^{-1} (C–H in-plane

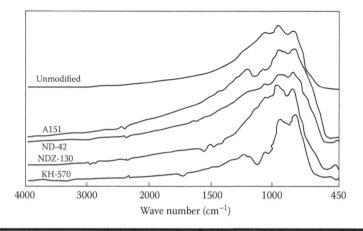

Figure 3.17 Infrared spectrum of zinc oxide whiskers before and after modification.

bending vibration), 1459 cm⁻¹ (stretching vibration of carbonyl), 1537 cm⁻¹, and 2000–3000 cm⁻¹ (stretching vibration of methyl) appear. When zinc oxide whiskers are modified with the titanate coupling agent ND-42, new adsorption peaks between 1400 and 1600 cm⁻¹ (deformation vibration of N–H and skeletal vibration of the C–C bond of benzene) appear. When zinc oxide whiskers are modified with the silane coupling agent KH-570, new infrared adsorption peaks at 1732 cm⁻¹ (stretching vibration of carbonyl) and 1126 cm⁻¹ (stretching vibration of C–O–C bond) appear. Furthermore, when zinc oxide whiskers are modified with the coupling agent A151, new infrared adsorption peaks at 1155 cm⁻¹ (stretching vibration of C–H bond) and 1410 cm⁻¹ (deformation vibration of C–H bond) appear. In conclusion, the appearance of new infrared adsorption peaks at 500 cm⁻¹ and 2300 cm⁻¹ indicates the formation of new chemical bonds between whiskers and coupling agents.[22]

Figure 3.18 shows infrared spectroscopic information of aluminum borate whiskers before and after being modified with sodium stearate. As shown in Figure 3.18, when aluminum borate whiskers are modified, the infrared adsorption peaks at 2918 cm⁻¹ (asymmetrical stretching vibration of C–H bond in CH_3) and 2851 cm⁻¹ (symmetrical stretching vibration

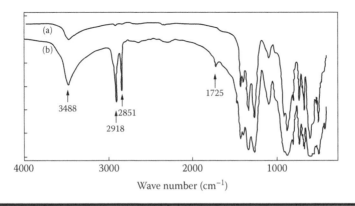

Figure 3.18 **Infrared spectrum of aluminum borate whiskers before and after modification. (a) Unmodified and (b) modified.**

of C–H bond) obviously increase. Moreover, the appearance of absorption peaks at 3488 cm^{-1} (carbonyl absorption peak) and 1725 cm^{-1} (ester carbonyl) indicates that ester bonds are formed between sodium stearate and the hydroxyl on the surface of aluminum borate whiskers and both chemical and physical absorptions exist between sodium stearate and whiskers.

3.5.1.5 Scanning Electron Microscopy and Energy Dispersive Spectrometry Analyses

The surface morphology of whiskers can be characterized and analyzed by scanning electron microscopy (SEM) and energy dispersive spectrometry (EDS). Figures 3.19 and 3.20 show SEM images and EDS spectrum of potassium titanate whiskers treated with polymethyl methacrylate (PMMA). According to the SEM results, the surface of unmodified whiskers is smooth without attachments, while the surface of modified whiskers is rough with attachments. The EDS spectrum of modified whiskers shows the characteristic peak of carbon (C). Both SEM and EDS results prove that the whisker surface is cladded with PMMA molecules.[35]

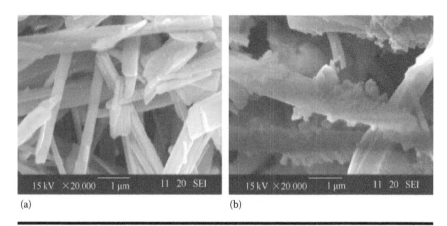

(a)   (b)

Figure 3.19 **SEM images of potassium titanate whiskers before and after modification. (a) Unmodified. (b) Modified.**

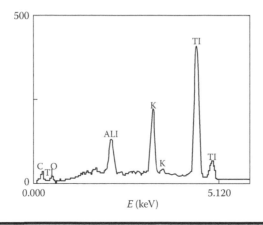

Figure 3.20   EDS spectrum of modified whiskers.

3.5.1.6 Thermal Gravimetric Analysis

The surface characteristics of modified whiskers can be analyzed by detecting the thermal weight loss curve. Figure 3.21 shows the thermal gravimetric analysis (TGA) curve of potassium titanate whiskers modified with PMMA.

As can be seen from the TGA curve in Figure 3.21, there is an obvious weight loss between 312.5 and 381.5°C (that is the weight loss area of PMMA). The calculated total weight loss

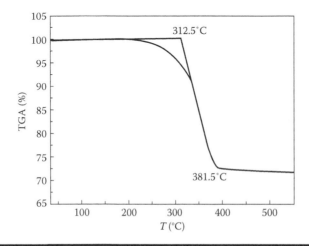

Figure 3.21   TGA curve of modified potassium titanate whiskers.

rate of modified whiskers is 28.33%, corresponding to a 39.5% weight addition, which is consistent with the theoretical value of 38.25%. Based on the preceding analysis, it is concluded that PMMA molecules are grafted onto the whisker surface.[35]

3.5.2 Indirect Method

In the indirect method, surface-modified whiskers are filled into polymers to prepare composite materials and the surface modification effects of whiskers can be evaluated by studying characteristics of composites, such as mechanical, thermal, and electrical properties. This method is the most often used.

For example, Yang et al.[36] compared the surface modification effects of several common coupling agents by determining the mechanical performance of polypropylene–nylon composites reinforced with modified potassium titanate whiskers. As shown in Tables 3.18 and 3.19, the surface modification effect of silane coupling agent is obvious. For a polypropylene system filled with potassium titanate whiskers, the total effect of KH-550 is the best. Whiskers modified with the silane coupling agent KH-560 is suitable for a nylon-66 system, which is indicated by the tensile strength increase of

Table 3.18 Whiskers Modified with Different Coupling Agents and Applied in Polypropylene

Sample	Tensile Strength (MPa)	Flexural Strength (MPa)	Notched Impact Strength (J/m)
Pure polypropylene	34.3	32.6	42.6
Unmodified whiskers	33.9	49.6	38.0
Modified with KH-550	39.5	55.1	62.3
Modified with KH-560	37.6	50.4	57.6
PP-g-MAH+KH-550	42.2	56.9	74.1

Table 3.19 Whiskers Modified with Different Coupling Agents and Applied in Nylon-66

Sample	Tensile Strength (MPa)	Flexural Strength (MPa)	Notched Impact Strength (J/m)
Pure nylon-66	71.6	114.0	37.4
Modified with KH-550	75.7	105.2	33.5
Modified with KH-560	80.8	118.2	43.3
Modified with NZD-201	73.1	64.5	34.8
Modified with NZD-401	73.6	98.8	31.5

12.85% and impact strength increase of 15.78%. The modified whiskers can suffer certain loads and locally resist stress in the surrounding matrix because of the improved surface combination of whiskers and resin, which gives composites great mechanical properties.

In a word, because inorganic whiskers have strong polarity, surface modification is needed to make whiskers well dispersed in weak polar polymers and give them excellent physical and chemical performances.

References

1. Junwei Li, Zhifeng Liu. Properties of HDPE composite filled with hollow glass beads. *Plastics*, 40(3):29–31, 2011.
2. Xianyong Chen, Qin Tang, Daijun Liu. Preparation and characterization of calcium carbonate micro-spheres. *Chemical Research and Application*, 24(2):314–317, 2012.
3. Minghe Sun, Qiuju Sun, Li Dai. Progress on application research of inorganic whiskers in polymer-based composites. *Fine and Specially Chemicals*, 18(9):9–13, 2010.
4. Zhiliang Jin, Zhihong Zhang, Wu Li. Progress of inorganic whisker materials research. *Sea-Lake Salt and Chemical Industry*, 31(5):4–13, 2002.

5. Wu Li. *Inorganic whisker.* Beijing: Chemical Industry Press, 2005.
6. Jing Wang, Litong Zhang, Laifei Cheng et al. Preparation and mechanical properties of silicon carbide whisker reinforced silicon carbide mini composites. *Journal of Aeronautical Materials,* 29(1):68–71, 2009.
7. Dong Jin. Preparation methods and application progress of inorganic whisker materials. *Fine Chemical Industrial Raw Materials and Intermediates,* (10):23–27, 2010.
8. Hua Ge, Xiuyue Du. Scanning electron microscope of inorganic whisker. *Journal of Chinese Electron Microscopy Society,* 25:365–366, 2006.
9. Litao Li. *Study and application of reinforced and wear-resistant whisker.* Wuxi: Jiangnan University, 2008.
10. Jiru Meng, Zhao Lei, Liang Guozheng et al. The applications of inorganic whiskers in polymers. *New Chemical Materials,* 29(12):1–6, 2001.
11. Takashi Sasaki, Kttagawa Takayki, Sato Shinataro et al. Core/shell and hollow polymeric capsules prepared from calcium carbonate whisker. *Polymer Journal,* 37(6):434–438, 2005.
12. Takashi Sasaki, Shoko Kawagoe, Hajime Mitsuya et al. Glass transition of crosslinked polystyrene shells formed on the surface of calcium carbonate whisker. *Journal of Polymer Science B: Polymer Physics,* 44(17):2475–2485, 2006.
13. Zhengguang Dang. The present situation of the application of inorganic whisker. *Liaoning Chemical Industry,* 36(11):777–7804, 2007.
14. Min Zou, Qilin Wang, Guangqiang Ma. Effect of heat treatment on mechanical performance of $CaSO_4$ whiskers/PP-R resin composites. *Transactions of Materials and Heat Treatment,* 30(2):49–52, 2009.
15. Jing Dai. *The synthesis and characterization of boric acid ester coupling agent and research of borate whisker modified polymer.* Qinghai: Qinghai Institute of Salt Lakes, Chinese Academy of Sciences, 2006.
16. Qiaoying Jia, Xiaoyan Ma, Guozheng Liang. Whiskers and their applications in polymer. *Polymer Bulletin,* 28(6):71–77, 2002.
17. Xiaoming Cui. Research and application progress of inorganic whisker. *Fine Chemical Industrial Raw Materials and Intermediates,* 16(5):25–30, 2007.

18. Yewen Cao, Jiachun Feng, Peiyi Wu. Simultaneously improving the toughness, flexural modulus and thermal performance of isotactic polypropylene by α-β crystalline transition and inorganic whisker reinforcement. *Polymer Engineering and Science*, 50(2):222–231, 2010.

19. Xuetao Luo, Lifu Chen, Qianjun Huang. Preparation and properties of PVDF matrix piezoelectric composites containing highly aligned $BaTiO_3$ whiskers. *Journal of Inorganic Materials*, 19(1):183–189, 2004.

20. Baofeng Pan. Study on properties of SMC whisker/polypropylene composite. *Plastics Manufacture*, 40(8):72–74, 2008.

21. Nanying Ning, Feng Luo, Ke Wang. Interfacial enhancement by shish-calabash crystal structure in polypropylene/inorganic whisker composites. *Polymer*, 50(15):3851–3856, 2009.

22. Yuexin Han, Lixia Li, Wanzhong Yin et al. On the surface modification of basic magnesium sulfate whiskers. *Journal of Northeastern University (Natural Science)*, 30(1):133–136, 2009.

23. Changsheng Li, Wenbin Mu, Jie Hu. Preparation and tribology capability of potassium titanate whiskers reinforced MC nylon. *Engineering Plastics Application*, 36(1):4–7, 2008.

24. Xiuyin Zhang, Wensheng Wu, Aifeng Tian. The effect of whisker fraction on flexural strength of whisker-PMMA composite reinforced with potassium titanate whiskers. *Journal of Clinical Stomatology*, 25(3):151–154, 2009.

25. Weiping Liu, Jianping Zhou, Keqiang Qiu. Surface modification and characterization of zinc oxide whiskers. *China Plastics Industry*, 32(5):47–49, 2004.

26. Mingshan Yang, Chenghan He. Preparation of calcium carbonate whiskers and modification of PP with it. *Modern Plastics Processing and Applications*, 20(6):21–24, 2008.

27. Xiaoli Wang, Yiming Zhu, Yuexin Han. On the surface modification of basic magnesium sulfate whiskers. *Journal of Northeastern University (Natural Science)*, 29(10):1494–1498, 2008.

28. Xiaolang Chen, Wangchun Lv, Wei Wang et al. Structure and morphology of PANI/T-ZnOw composite. *China Plastics Industry*, 36(7):50–52, 2008.

29. Cuifeng Wan, Shengming Jin. Surface modification of T-ZnO whisker and its application in antistatic epoxy resin. *Materials Review*, 21(5):375–377, 2007.

30. Wenyun Zhang, Yanbo Yuan, Qinghua Chen et al. Influence of nano-silica content on flexural properties of the aluminum borate whisker and silica filler composites resins. *West China Journal of Stomatology*, 29(4):195–199, 2011.
31. Mingbo Guo, Hong Xu, Hongchen Gu. Surface treatment of inorganic whisker. CP200110015677, 9.
32. Changsong Wang, Xin Feng, Xiaohua Lu. Surface modification of potassium titanate whiskers with *n*-octadecyltrichlorosilane, *Acta Physico-Chimica Sinica*, 21(6):586–590, 2005.
33. Guisheng Zeng, Jianping Zou, Qiang Peng et al. Surface modification of T-ZnOw with silane coupling agent KH-570. *Journal of Functional Materials*, 41(3):410–413, 2010.
34. Min Zou, Qilin Wang, Guangqiang Ma. Effect of heat treatment on mechanical performance of $CaSO_4$ whiskers/PP-R resin composites. *Transactions of Materials and Heat Treatment*, 30(2):49–52, 2009.
35. Weiping Chen. *The study of the surface modification and characterization and interface property of potassium titanate whiskers*. Nanjing: Nanjing University of Science and Technology, 2004.
36. Ning Yang, Dayong Gui, Jiping Liu. Surface treatment of whisker with silane coupling agent and its applications. *Plastics Science and Technology*, 160(2):14–17, 2004.

Chapter 4

Application of Filling Inorganic Whiskers into Polymers

Inorganic whiskers as single crystal fibers have become a remarkable new type of filling material because of its high-intensity, heat-resistant, anti-friction, anti-corrosion, and fire-resistant properties and other special features.

Inorganic whiskers were first used in ceramic matrix and metal matrix composites. Ceramic matrixes possess special properties such as high-intensity, anti-friction, high-temperature-resistant, anticorrosion properties, and so on; however, their brittleness have adverse effects. Among various toughening methods, whisker toughening is one of the best methods and the toughening mechanism has been deeply studied. Current research focuses on applications in polymer matrix composites. Applying inorganic whiskers in polymer matrix composites has become a hot topic, which could combine the inflexibility, dimensional stabilization, and thermal stabilization with the ductibility of polymers to possibly produce materials needed in high technology and expand the application fields of existing polymer materials.

4.1 Composition of Polymer Composites Filled with Inorganic Whiskers

A polymer composite system filled with inorganic whiskers includes whiskers, matrix resin, coupling agent, or other surface treatment agent. Sometimes an antioxidant, thermal stability agent, plasticizer, crosslinking agent, dispersing agent, and other additives are also added if needed. Only matrix resin and additives are considered in this chapter because inorganic whiskers, coupling agents, and surface treatment agents and so on were discussed in detail in Chapters 2 and 3.

4.1.1 Matrix Resin

A matrix resin is an essential component in a filling system. It usually is the main component if present as a larger mass fraction, but it could be a secondary component if present as a smaller mass fraction. For example, packing material can account for 70%–80% in the preparation of a filling masterbatch. In both situations, the physicochemical properties of the matrix resin are significant for the performances of composite materials. Therefore, the matrix resin should satisfy the following demands.[1-3]

4.1.1.1 Requirements

4.1.1.1.1 Excellent Comprehensive Properties

The matrix resin should have excellent comprehensive properties, including mechanical, electrical, thermal performance, chemical resistance, anti-aging, flame retardancy, and so on. However, it is impossible for an individual matrix resin to have all the properties simultaneously, so the appropriate one should be chosen according to usage requirements and characteristics of packing materials. For example, fire retardancy, compatibility of filler and matrix resin, and dispersibility of filler in the resin should be taken into account to make the filled composite material meet usage requirements.

4.1.1.1.2 Strong Adhesive Force to Fillers

The important role of resin in the filling system is to combine all the fillers as a whole as a continuous phase, producing a new material with a new structure and properties. This adhesive action is very important. Whiskers cannot be used as a bearing material because they are short fibrous fillers with draw ratios ranging from 5 to 1000. However, they can greatly improve the mechanical properties of materials when bonded together as a whole. In addition, excellent adhesion with polymers can protect the filler from etching and erosion of environmental media, which would promote effective function of the fillers.

4.1.1.1.3 Outstanding Processing Capability

Easier molding processing conditions are desirable to reduce equipment investment, simplify manufacturing operations, and develop products with huge dimensions or complex geometric configurations. Processing properties include the following.

■ Resin with the proper fluidity should be chosen. Resin with low fluidity cannot adequately cover fillers and resists molding, whereas resin with high fluidity can be easily lost during molding, resulting in underfilled product and inadequate control of the resin to filler ratio.
■ The molding shrinkage of resin should be small, and the smaller the difference with packing shrinkage, the better it is. If not, large shrinkage stress at the packing–matrix resin interface is easy to be generated, which affects the intensity of the filling and the dimensional stability.
■ Thermosetting resins should have a proper curing time. A prolonged curing time will affect production efficiency, whereas a short curing time is difficult for processing and application of large products.

At present, matrix resins used for studying filling modification with inorganic whiskers include thermoplastic and thermosetting resins. Thermoplastic resins include polyethylene, polypropylene, poly(vinyl chloride), polytetrafluoroethylene, polystyrene, polyamide resins, and so forth. Thermosetting resins contain epoxy resin, unsaturated polyester resin, and so on.

4.1.1.2 Thermoplastic Resins

4.1.1.2.1 Polyethylene

Polyethylene (PE) is formed by the polymerization of ethylene, and the formula is

$$\left(\!\!\begin{array}{c} CH_2CH_2 \end{array}\!\!\right)_{\!n}$$

PE is nontoxic and has an excellent electrical insulating property, chemical resistance, processing fluidity, outstanding mechanical properties, low temperature resistance, and light transmission. Therefore, the production and development speeds of PE and its products are very fast. PE can be divided into four categories.

1. Low-density polyethylene (LDPE) is usually produced via a high-pressure (147.17–196.2 MPa) method, and thus is also called high-pressure polyethylene. A PE molecular chain produced by the high-pressure method contains more branches (an average of about 21 branches per 1000 carbon atoms). Therefore, high-pressure PE has a low degree of crystallinity (45%–65%), low density (0.910–0.925 g/cm^3), light weight, softness, cold resistance, and good impact resistance, and is suitable for production of films.
2. High-density polyethylene (HDPE) is produced via a low-pressure method and is also called low-pressure polyethylene. HDPE has fewer branched-chain molecules, a high degree of crystallinity (85%–95%); high density (0.941–0.965 g/cm^3); and high usage temperature, hardness, mechanical strength, and chemical resistance. It is suitable for the production of corrosion-resistant parts and insulating parts.
3. Linear low-density polyethylene (LLDPE) is a copolymer of ethylene and α-olefin. The synthesis method is

basically the same as for HDPE, and it therefore has a straight-chain molecular structure. With the addition of an α-olefin monomer, many short and regular chain branches are produced. In the molecular chain structure, the regularity, density, and crystallinity lie between those of LDPE and HDPE but are closer to those of LDPE.

4. Metallocene polyethylene (MPE) is a new PE product initiated and polymerized by metallocene calalyst. It has a narrow molecular weight distribution ($d = 2$); uniform molecular chain structure; higher crystallinity, strength, toughness, and transparency; and excellent performance. However, MPE melt has a high viscosity that makes processing difficult, which limits its application.

4.1.1.2.2 Polypropylene

Polypropylene (PP) is formed by polymerization of propylene. The formula is

$$\left(\text{CHCH}_2\right)_n$$
$$\text{CH}_3$$

PP can be divided into atactic, isotactic, and syndiotactic PP based on its molecular structure. Industrial isotactic PP is a colorless and odorless solid, with a density of 0.90–0.91 g/cm³. It possesses good heat and chemical resistance and toughness, and the usage temperature ranges from –30 to 140°C. The disadvantages are poor impact, temperature, aging resistance, and surface printing performance, as well as large molding shrinkage. It is used mainly in the manufacture of plastic products, such as home appliances, home appliance parts, packaging films, strapping material, sterilized medical utensils, and also fiber products (carpets, etc.).

When isotactic PP is produced, a small amount of atactic PP appears as a by-product that is a noncrystalline waxy solid with low microstrip viscosity at room temperature. The relative molecular mass is typically only 3000–10,000. It has no single use value; however, it can be used as a carrier resin and dispersant in the manufacture of PP filling masterbatch.

4.1.1.2.3 Poly(vinyl Chloride) Resin

Poly(vinyl chloride) (PVC) resin is formed by the polymerization of vinyl chloride. The structure is

$$\left(\!-CHCH_2\!-\!\right)_n$$
$$|$$
$$Cl$$

PVC has good electrical insulating properties and good chemical resistance (resistant to concentrated hydrochloric acid, 90% sulfuric acid, 60% nitrate, and 30% sodium hydroxide), but because of poor thermal stability it is necessary to add a thermal stabilizer during PVC molding. PVC easily absorbs plasticizers. With an increasing amount of plasticizer, PVC changes from hard to soft, and thus there are hard and soft PVC products. However, because of their high glass transition temperature (T_g), hard products cannot withstand low temperature.

Because of poor thermal stability, PVC can decompose with long heating times, releasing chlorine hydride gas, which changes the color of PVC; thus PVC has a narrow application range, with operation temperatures ranging from –15 to 55°C. PVC is flame retardant and becomes self-extinguishing if ignited. Its transparency is better than that of PP and PE. PVC is highly versatile, with soft products used for agricultural films, artificial leather, and so forth; applications for the hard products include industrial anti-corrosion and structural materials, building materials, as well as thin insulation layers of wires and cables.

4.1.1.2.4 Polystyrene Resin

Polystyrene (PS) is formed via the polymerization of styrene. The structure is

The macromer backbone of PS is a saturated hydrocarbon chain with phenyl side groups, and it is difficult to form an ordered structure because of its asymmetric molecular structure and huge volume of benzene, which gives the benzene ring greater rigidity. Therefore, PS is a typical amorphous thermoplastic resin. PS is a colorless transparent thermoplastic plastic with a hard and brittle texture, easy dyeability, and good processing flowability. Therefore, it is particularly suitable for injection molding of various commodities, toys, instrument cases, automatic wire lamp shade frames, disposable tableware, and packaging and thermal insulation of shock-proof materials.

PS is typically a hard and brittle material. It has relatively poor chemical stability and thus can be dissolved in many organic solvents and can be corroded by highly concentrated acid and alkali. PS is not grease resistant and easily changes color after exposure to ultraviolet (UV) radiation. It also has poor heat resistance, with the highest temperature at continuous use of 60 to 80°C.

4.1.1.2.5 Polytetrafluoroethylene Resin

Polytetrafluoroethylene (PTFE) resin is formed via the polymerization of polytetrafluoroethylene. The structure is

$$\left(CF_2CF_2 \right)_n$$

The trade name of PTFE is Teflon®. PTFE has excellent resistance to both high and low temperatures. The melting point is 327°C and the long-time use temperature is between −200 and 260°C. Its chemical resistance is superior to that of all other plastics, and it is insoluble in strong acid and alkali as well as organic solvents; it is unchangeable even in boiling nitrohydrochloric acid. It is one of the best corrosion-resistant materials in the world today and widely used for various situations requiring anti-acid, anti-alkaline, and anti-organic solvents. In addition, PTFE does not absorb moisture, is noncombustible, and is very stable to oxygen and UV radiation and therefore has excellent weathering properties. PTFE also possesses tightness, high lubrication without viscosity, electrical insulation, and good anti-aging characteristics.

As engineering plastic, PTFE can be made into Teflon tubes, sticks, belts, boards, films, and generally used to make corrosion-resistant pipes, containers, pumps, and valves with high performance requirements as well as guidance radar, high-frequency communications equipment, radio equipment, etc. Its disadvantage is that its flow is difficult even when PTFE is heated to above the melting point, which makes PTFE hard to mold, and only the sintering molding method can be used, similarly to powder metallurgy.

4.1.1.2.6 Polyamide

Polyamide (PA) is commonly known as nylon, which is a general term for polymers containing many repeating amide groups on the main molecular chain. It is formed by polycondensation of principally dicarboxylic acids and diamines, or amino acids. Common varieties of nylon are nylon-6, nylon-66, nylon-1010, and so on. PAs are usually white to pale yellow nontransparent solids, with melting points of 180–280°C and a density of 1.05–1.15 g/cm^3. PA plastic was the first engineering plastic developed, and its production is the largest among all engineering plastics.

PA has properties such as wear resistance, toughness, light weight, chemical resistance, easy dyeability, oil resistance, heat and cold resistance, ease of formation; it is also self-lubricating, nontoxic, odorless, mildew free, and self-extinguishing. It is not soluble in ethanol, acetone, ethyl acetate, and common hydrocarbon solvents. But PA is soluble in phenolic, sulfuric acid, methane acid, acetic acid, and certain inorganic salt solutions. Because of its high mechanical strength, good wear resistance, and self-lubricating properties, PA is used mainly in the production of synthetic fibers and reinforced plastics. However, it absorbs water vapor easily and has poor creep resistance, and therefore is not suitable for the production of electrical insulating materials and precision parts.

4.1.1.3 Thermosetting Resins

4.1.1.3.1 Epoxy Resin

Epoxy (EP) resin contains epoxy groups in its molecular structure. There are many types of EP resin, but more than 90% of EP is bisphenol A-type epoxy resin, which is formed via the polymerization of bisphenol A and epichlorohydrin. The molecular structure is

$$CH_2-CH-CH_2-\!\!\left(\!O-\!\!\bigcirc\!\!-\!\!\overset{\overset{\displaystyle CH_3}{|}}{\underset{\underset{\displaystyle CH_3}{|}}{C}}-\!\!\bigcirc\!\!-O-CH_2-CH-CH_2\!\!\right)_{\!\!n}\!\!-O-\!\!\bigcirc\!\!-\overset{\overset{\displaystyle CH_3}{|}}{\underset{\underset{\displaystyle CH_3}{|}}{C}}-\!\!\bigcirc\!\!-O-CH_2-CH-CH_2$$

An EP resin with low molecular weight is a yellow or amber transparent liquid with high viscosity; an EP resin with high molecular weight is an odorless and tasteless solid. It is soluble in acetone, cyclohexanone, glycol, toluene, ethylene, styrene, and so forth. EP resins are mainly made into liquid products with various viscosities because of different molecular weights. The difference of molecular weight can be represented by EP value: the larger the molecular weight, the smaller the EP value.

EP resin can be crosslinked and cured into a network structure solid under treatment with amine and acid curing agents. No low molecular weight product is produced during the curing process; therefore, products can be low-pressure molded with low shrinkage and no pores, which makes EP resin a thermosetting resin. EP resin has high bond strength and can be cured at low temperature and normal pressure, and is therefore widely used in adhesives of metallic and nonmetallic (ceramics, glass, timber, etc.) materials. Glass fiber-reinforced plastic products made of EP resin reinforced by glass fiber and its fabric have outstanding mechanical and electrical insulation properties, and major products include electrical switching devices, dashboards, printed circuit boards, pressure vessels, chemical pipelines, and mechanical parts.

4.1.1.3.2 Unsaturated Polyester Resin

Unsaturated polyester (UP) resin is an unsaturated linear polyester resin polycondensed by unsaturated dibasic acid or anhydride (mainly maleic anhydride and fumaric acid), a certain amount of saturated dibasic acids (such as phthalic acid and terephthalic acid), and dihydric alcohol or polyhydric alcohols (such as ethylene glycol, propylene glycol, and glycerin).

UP has high tensile, bending, and compression strengths. It has good resistance to water and dilute acids and bases, but poor resistance to organic solvents. UP has differing resistances to chemical corrosion depending on the chemical structure and geometry. The heat distortion temperature of the majority of UP resins is 50°C–60°C. Some resins with good heat resistance have heat distortion temperatures that can be as high as 120°C, and these are mainly used for glass fiber-reinforced plastics, paint, daub, molding powder, and so on.

The main chain of a UP contains ester bonds and unsaturated double bonds, and both ends of the macromolecular chain have acid groups and hydroxyl groups. The double bonds in the main chain and vinyl monomers (such as styrene) can have crosslinking copolymerization, which changes the state of UP from fusible and soluble into infusible and insoluble. The ester bonds in the main chain can hydrolyze, which can be accelerated by an acid or base. However, if the ester bonds are crosslink copolymerized with styrene, it can greatly reduce the hydrolysis reaction. Furthermore, the carboxyl groups at the chain end can react with alkaline earth metal oxides or hydroxides (such as MgO, CaO, $Ca(OH)_2$, etc.), which can extend the UP molecular chain. The extension of the molecular chain enhances the viscosity of the resin in a short surge until it becomes a nonflowing, nonsticky, and gel-like material. Resin in this state is not crosslinked and therefore can be dissolved in suitable solvents. It also has good liquidity when heated and large volume shrinkage during curing.

4.1.1.3.3 Polyether Ether Ketones

Polyether ether ketones (PEEKs) are a polycondensation product of 4, 4′-difluorophenyl ketone, hydroquinone, and anhydrous sodium carbonate (or potassium carbonate). The molecular structure is as follows:

The benzene and benzophenone structure makes the acromolecular chains rigid, and ether linkages make the macromolecular chains flexible. Thus, PEEK is an engineering thermoplastic with both rigidity and flexibility.

The most important advantage of PEEK is its heat resistance. The melting point is 334°C, the highest continuous use temperature is 240°C, which can reach 300°C if PEEKs are reinforced by glass fiber or carbon fiber. Furthermore, PEEKs have good mechanical and electrical insulation properties, thermal oxidation stability, radiation resistance, and corrosion resistance. PEEKs are used mainly in the military and space fields and in the oil, machinery, and chemical industries for products such as printed circuit boards, electrical insulating materials, structural parts on planes, machinery parts, valves, anti-corrosion coating, and so on. The disadvantage is harsh processing conditions.

4.1.2 Additives

Under normal conditions, some resins such as PE, PS, and PA can be directly molded; however, to meet the requirements of a particular application performance, treatments such as coloring, filling, anti-aging, anti-flaming, foam, and so on are needed. Therefore, preparation of almost every resin involves coordination with additives.[4]

Additives not only can improve the processability of resin and compensate for the polymer material itself, but also can endow products with a variety of valuable performances. They can even transform polymer materials with low practical value into highly valuable materials. It is not an exaggeration to say that there would be no development of the plastics industry without plastic additives.

In general, the concentration of the additive is represented by the mass fraction of additive per hundred mass fraction of resin, expressed as "parts per hundred resin (PHR)." The concentration of most additives is small, from a few tenths to several parts, some up to ten parts or more parts.

Large varieties of plastic additives are available: inorganics and organics, low molecular weight compounds and polymers, and single items and mixtures. With the extended application of plastics, new additives are emerging and it is difficult to summarize them fully. Several commonly used additives are described in the following sections.

4.1.2.1 Antioxidants

Most polymers can react with oxygen, and the oxidation reaction is faster especially during thermal processing or in sunlight, which shortens the working life of products. Antioxidants are the additives most widely used to inhibit or delay the degradation of polymers by the action of oxygen or ozone in the atmosphere.

Antioxidants have two kinds of mechanisms of actions: inhibiting free radical chain reactions and decomposing hydroperoxide. Free radical inhibitors are primary antioxidants, including amines and phenols; hydroperoxide decomposers are known as secondary antioxidants, including phosphites and thioesters, which are usually used with primary antioxidants.

Aromatic amine antioxidants are also known as rubber anti-aging agents; they have the largest production quantity and are used mainly in rubber products. These antioxidants have low prices and remarkable antioxidant effects; however, they could change the color of products, which limits their application in light colored and white goods. Significant aromatic amine antioxidants are diphenylamine, *p*-phenylenediamine, quinoline, and their derivatives or polymers, and they can be used in natural rubber, styrene butadiene rubber (SBR), chloroprene rubber, and isoprene rubber.

Hindered phenol antioxidants are some of the phenolic compounds with sterical hindrance. They have remarkable resistance to thermooxidation effects and have no pollution effects on products; they therefore develop rapidly. These antioxidants have many varieties, and the important products include 2, 6 *tert*-butyl-4-methylphenol, bi-(3, 5 *tert*-butyl-4 hydroxyphenyl) thioether, and β-(3, 5-*tert*-butyl-4-hydrophenyl) propionate pentaerythritol tetraester. Such antioxidants are mainly used in plastics, synthetic fibers, latex, petroleum products, food, drugs, and cosmetics. The structure of hindered phenolic antioxidant is shown in Figure 4.1.

Single phenol and bisphenol antioxidants, such as butylated hydroxytoluene (BHT), 2246, and bisphenol A, because of their lower molecular weight and high volatility and mobility, can color plastics easily; therefore their use in plastics has recently declined greatly. Polyphenolic antioxidants 1010 and 1070 are the leading plastic antioxidant products in the world. 1010 is the best plastic antioxidant because of its high

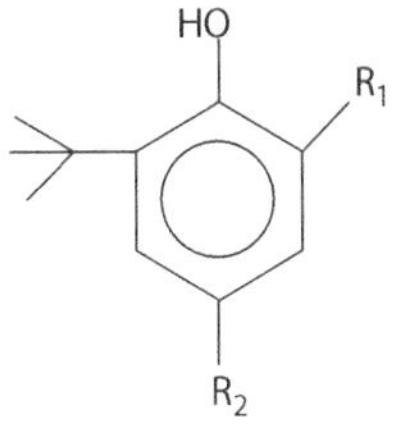

R_1:-CH$_3$,-CH$_2$-,-S-; R_2:R, -CH$_2$CH$_2$COOR

Figure 4.1 Structure of a hindered phenolic antioxidant.

molecular weight, good compatibility with plastic materials, excellent antioxidant effect, and largest consumption.

Secondary antioxidants, such as thiodipropionate diester, are often used together with hindered phenolic antioxidants and the effect is remarkable. The main products are dodecanol double ester, bis-ester, and bis-tetradecene octadeca. Phosphite ester is also a secondary antioxidant and the main products are trioctylcitrate, tricaprylate, 3-pentanediol monoisobutyrate, and hexadecyl ester, such as antioxidants 168, 626, and 618. The production and consumption of 168 are second only to those of antioxidant 1010. Most manufacturers that produce 1010 and 1070 also produce 168. 626 and 618 are mainly used to produce plastic material or products processed at high temperatures of about 300°C, with the effect of improving the ability of antithermal oxidation and maintaining a good appearance of plastic products.

Different types of primary and secondary antioxidants, or the same types of antioxidants with different molecular structures, have different functions and application effects and both have their own strengths. A composite antioxidant is compounded with two or more different types of antioxidants or with different varieties of the same type that are mutually complementary, thereby achieving optimum antithermal oxidation and aging effects with minimal amount of antioxidants and lowest cost.

4.1.2.2 Light Stabilizers

Under sunlight or strong fluorescence, plastic, rubber, and synthetic fiber products will absorb UV rays and cause self-oxidization, leading to the degradation of the polymer, which deteriorates the appearance and mechanical properties of the products. This process is called photoredox reaction or photoaging. Short-wavelength UV radiation is an extremely important cause of outdoor aging of polymers. A light stabilizer, also known as a UV protectant, is an additive that can inhibit photooxidation and prolong the outdoor working life of polymers.

The mechanisms of light stabilizers are different because of their structures and different species. Some can shield and reflect UV rays or absorb them and turn them into harmless thermal energy; some can quench molecules triggered by UV rays or excited states of groups and return them to the base state, therefore eliminating or reducing the possibility of a photoredox reaction; some can catch free radicals generated by photoredox, thereby preventing free radical reactions that can cause product aging and the ensuing damage.

Light stabilizers can be divided into four categories.

1. Light shielding agents are capable of reflecting or absorbing UV rays of sunlight, that is, setting a barrier between the light source and polymer, thus preventing UV absorption into the polymer interior and protecting the polymer. Light stabilizers mainly include carbon black, zinc oxide, titanium dioxide, and zinc barium.

2. UV absorbers are a class of organic compounds with high lightfastness that could selectively strongly absorb UV rays from sunlight that are harmful to polymers. The most widely used UV absorbers in industry are benzophenone, salicylic acid, and benzotriazole.

3. A quencher is a kind of light stabilizer that could effectively transfer the excited state energy of photosensitive groups in polymers and dissipate them in a harmless form, thus preventing light degradation reactions of polymers. The main quenchers are metal complexes, such as the organic complexes of nickel, cobalt, and iron.

4. A radical scavenger acts by capturing and eliminating free radicals to stop the autooxidation chain reaction, and mainly includes hindered amine derivatives.

The following factors should be considered when choosing a light stabilizer.

1. It can effectively absorb UV rays with wavelengths of 290–400 nm, can quench the energy of excited state molecules, or can capture free radicals.
2. It has good light and thermal stability.
3. It has good compatibility and no leakage during use.
4. It is resistant to hydrolyzation and to extraction from water and other solvents.
5. It has low volatility and is low polluting.
6. It is nontoxic or has only low toxicity, is inexpensive, and can be obtained easily.

4.1.2.3 Heat Stabilizers

A heat stabilizer is an additive that could prevent or reduce the degradation or crosslink of polymers caused by heat during use and prolong the working life of composite polymers. It is a special additive for PVC–vinyl chloride copolymers and poly(vinylidene chloride), mainly used in PVC. PVC and other chlorine-containing resins are prone to degradation and release hydrogen chloride (HCl) during processing and heating because of their unstable structure, which leads to thermal instability and therefore causes a great deal of harm. Therefore, heat stabilizers are especially important for PVC and other chlorine-containing resins. They include primary stabilizers such as metal salts of fatty acids and organic tin compounds and auxiliary stabilizers such as epoxy compounds, phosphite ester, and polyalcohol.

According to their major components, common heat stabilizers can be divided into base heat stabilizers, fatty acids, organic tin compounds, composite heat stabilizers, and pure organic compounds.

1. Base heat stabilizers are inorganic and organic acid lead salts combined with a base. They have excellent heat resistance, weathering resistance, and electrical insulating properties, as well as low cost, poor transparency, and a certain toxicity; the concentration generally is 0.5%–5.0%.
2. Fatty acid heat stabilizers are compounds consisting of fatty acids and metal ions and are also known as metal soap heat stabilizers. Their performances depend on the types of metal ions and acid radical groups. The general concentration is 0.1%–3.0%.
3. Organic tin heat stabilizers could react with unstable chlorine atoms in PVC molecules and form ligands, and the carboxylic ester group in organotin could substitute for the unstable chlorine atoms in ligands. Features of such heat stabilizers are high stability, good transparency, and excellent heat resistance. The disadvantage is a relatively high price.
4. Composite heat stabilizers are liquid or solid compounds based on salts or metal soaps and compounds based on organic tin. Metal salts include Ca–Mg–Zn, Ba–Ca–Zn, Ba–Zn, and Ba–Cd; common organic acids include organic fatty acids, naphthenic acid, oleic acid, benzoic acid, and salicylic acid.
5. Organic compound heat stabilizers include not only primary stabilizers that can be used alone (mainly organic nitrogen compounds), but also polyalcohols and phosphite esters with a high boiling point. Phosphite esters are often used together with metal stabilizers to enhance the weather resistance and transparency of composite materials and improve surface color and luster of products.

4.1.2.4 Plasticizer

Any additive that is added into the polymer system and can increase the plasticity of the system can be called a plasticizer. A plasticizer usually is a solvent-like matter that is added to a plastic resin or elastomer material to improve its plasticity, flexibility, stretchability, or expansibility.

Plasticizers can be divided into two types based on their mode of action: interior and exterior plasticizers. They can also be divided into general and special plasticizers according to their functions. A general plasticizer has broad applicability and no special features, for example, phthalate esters. A special plasticizer has some special features in addition to the functions of general plasticizers, such as cold tolerance, flame retardance, environmental protection, and permanence.

A plasticizer is one of the important additives extensively used for PVC, and its main role is to soften products, improve their resistance to low temperature, and make them easy to process. Plasticizers are generally liquid organics with a high boiling point, and few are solid with a low melting point and rubber-type polymers. The vast majority of plasticizers are phthalate esters; others are phosphate esters, dibasic acid esters of fatty acids, and chlorinated hydrocarbons. At present, 80% of plastics enterprises in China commonly use dioctyl-phthalate (DOP), dibutyl phthalate (DBP), and phthalate isono-nylphthalate (DINP), which are still widely used in soft articles such as PVC hose, thin film, man-made leather, and so forth.

With the development of the PVC industry, the plasticizer industry is growing steadily, and phthalate esters plasticizers continue to take first place. Di(2-ethylhexyl)phthalate (DEHP) has been controversial for many years because of its toxicity, but it is still the major plasticizer of PVC with a rapidly growing market demand because of its good overall performance, excellent practicability, ideal processing performance, low price, and available raw materials.

4.1.2.5 *Crosslinking Agent and Auxiliary Crosslinking Agent*

A crosslinking agent causes a crosslinking reaction to occur in a polymer material; the additive that is used to improve crosslinking efficiency is known as an auxiliary crosslinking agent.

Crosslinking methods used for polymers include physical and chemical crosslinking, and a crosslinking agent is used for chemical crosslinking. Crosslinking interconnects polymer chains, enhances intermolecular forces, and improves the mechanical properties of materials. Organic peroxide is the most widely used crosslinking agent. To improve the degree of crosslinking of resin and change some physical and mechanical properties, an auxiliary crosslinking agent is often used with organic peroxide; for example, *p*-divinyl benzene and poly(butadiene-1.2) are often used together with dicumyl peroxide.

A curing agent is used for curing liquid thermosetting resin, and essentially is a special crosslinking agent. Curing agents can create new bonds between linear molecules, changing them to a three-dimensional structure thus improving the mechanical properties of resin, such as impact strength, tensile strength, modulus, creep resistance, and so forth. Common curing agents used for epoxy resin are amines, anhydrides, and low molecular weight polyamide; other thermosetting resins, such as phenol resin, unsaturated polyester, alkyd resin, and urea-formaldehyde resin, could all use a corresponding curing agent to accelerate the speed of curing.

4.2 Preparation Methods of Composite Material Filled with Inorganic Whiskers

Performances of materials not only depend on raw materials and recipes, but also are closely related to processing methods. An improper molding method not only cannot result in products with expected performances, but also can destroy the performance of raw materials. For example, excessively high processing temperatures will cause decomposition, crosslinking, and even coking of materials, thereby destroying the performance of raw materials. Therefore, the molding process is very important in the preparation of polymers.

The processing of polymers is carried out mostly in the liquid state, that is, in a melt or solution, such as hot press molding, extrusion, injection molding, rolling, blow molding, and so on. A few are sinter molded using solid powder, and a few are casting molded or reaction molded with monomers or prepolymers. Therefore, preparation methods of composite materials filled with inorganic whiskers are similar to molding methods for polymers, and are summarized in the following sections.[3,5]

4.2.1 Blending Dispersion Method

The blending dispersion method refers to the preparation of modified inorganic whiskers through surface modification of whiskers by using a coupling agent or surface modifier. Then the modified inorganic whiskers are placed into a matrix resin and mixed for a certain amount of time, followed by machine shaping. This method is similar to ordinary blending modification of polymers and industrial production is easily achievable. Most of the studies of filling inorganic whiskers use this approach.

4.2.1.1 Blend Extrusion Method

In this method, whiskers and resins are added to a twin-screw or single-screw blending extruder; with the help of a screw extrusion effect, the heated molten material is forced through the extrusion die under pressure and acquires a continuous profile with a constant cross section. The extrudates can also be molded directly, or granulated after extrusion to produce aggregates of composite materials, and the aggregates can be injection molded or compression molded. Commonly used extruders are a single-screw and twin-screw extruder. The twin-screw extruder has three ways of cycling: in the same direction, in the opposite direction, and at high speed in the same direction. Figure 4.2 shows a working schematic diagram of a single-screw extruder to produce tubes. Figure 4.3 is a schematic diagram showing the screw coordination of a twin-screw extruder.

Table 4.1 compares the performances of a single-screw and a twin-screw extruder.

In short, the twin-screw works better than the single-screw extruder in conveying, dispersive mixing, self-cleaning, and energy efficiency. A twin-screw extruder with the opposite direction has small shearing action and thus can be used for molding material under lower temperatures. A high-speed extrusion machine has significant mixing and blending effects

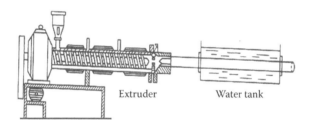

Extruder   Water tank

Figure 4.2   Schematic diagram of single-screw extruder to produce tubes to a water tank.

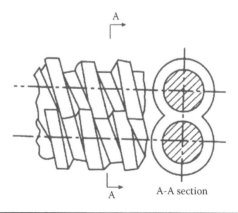

A-A section

Figure 4.3   Schematic diagram showing the screw coordination of a twin-screw extruder.

Table 4.1 Performance Comparison of Single- and Twin-Screw Extruders

Items	Single-Screw Extruder	Co-Rotating Twin Screw Extruder		Counter-Rotating Twin-Screw Extruder
		Low Speed	High Speed	
Transport efficiency	Small	Middle	Middle	Large
Dispersion-mixing	Small	Middle	Large	Large
Shearing action	Large	Middle	Large	Small
Self-cleaning	Small	Middle–large	Large	Middle
Energy utilization efficiency	Small	Middle–large	Middle–large	Large
Heat generation of machine	Large	Middle	Large	Small

Source: Zhihua Wu, Qi Yang. *Molding technology of polymer material.* Chengdu: Sichuan University, 2010.

and is used mainly in composite materials that are difficult to plasticize.

Because inorganic whiskers and some additives are powders, and most resins are granular, these materials are often premixed in a high-speed kneading machine to ensure the uniform mixing of whiskers and resins, which enables the powder particles to wrap evenly on the surface of polymer granules. In addition, because a single-screw extruder has worse mixing ability than a twin-screw extruder, whiskers and matrix resins and additives especially need to be premixed before extrusion. Figure 4.4 is a sketch of two common kneading machines.

Consider, for example, the preparation of polypropylene composite material filled with calcium sulfate whiskers

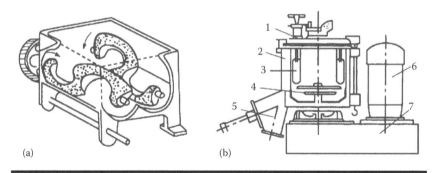

Figure 4.4 Sketch of two common kneading machines. (a) Z-type kneading machine. (b) High-speed kneading machine. 1, rotation container lid; 2, rotation container; 3, baffle; 4, high-speed impeller; 5, outlet of valve; 6, motor; 7, motor base.

in a single-screw extruder.[6] The operations are as follows. Accurately measure the amount of PP, surface-modified whiskers, toughening agent (ethylene–octene copolymer), and other additives and premix them in a high-speed kneading machine. Then, the evenly mixed material is extruded using a single-screw extruder, in zone I, 40–150°C; zone II, 160–180°C; zone III, 180–200°C, with a speed of 870 rpm, then is cooled, sectioned, and cut, and a composite aggregate of PP and calcium sulfate whiskers is obtained.

In the blending process, whiskers may become damaged or even broken because of the impact and collision between whiskers and resin particles and the shear caused by the rotation of the screw. The constant length-to-diameter (L/D) ratio of whiskers is the key for filling modification, so the speed of extrusion, the temperatures of different sections, and the position of the feed inlet of whiskers greatly influence the performance of the composite materials. Ge[7] adds whiskers from the feed inlet of the twin-screw extruder and the first outlet respectively, and the performances of the composite material are shown in Table 4.2.

From Table 4.2 we can see that the strengths of composite materials prepared with different feeding positions are quite different, especially the impact strength, which has an

Table 4.2 Influence of Different Feeding Positions to PP Performances Filled by Whiskers

Feed Position	Tensile Strength (MPa)	Elongation at Break (%)	Bending Strength (Mpa)	Impact Strength (KJ/m²)	Melt Flow Rate (g/10 min)
Feed inlet	30.6	3.9	42.7	10.3	4.7
The first exhaust exit	39.5	7.5	66.5	85.3	4.4

eightfold difference. Whiskers and polypropylene are mixed together and loaded from the feed inlet; the impact strength of the composite material is severely lost because of the damage of whiskers during the conveying process, which results in the severe decline of the L/D ratio of whiskers. However, when whiskers are added from the first outlet, because PP is basically melted when it arrives at the first outlet, whiskers are wrapped immediately when they enter the material hopper, and their reinforcing and toughening effects on PP are therefore achieved as a result of the maintenance of a good L/D ratio of whiskers.

In addition, single-screw and double-screw extruders affect the performance of composites filled by whiskers differently, as shown in Table 4.3.

From Table 4.3 we can see that under the same proportion of added whiskers, when calcium sulfate whisker modified PP is blended and granulated using twin-screw extrusion equipment, the tensile strength and bending strength of the composite material are higher than those of the material made with single-screw equipment; the impact strengths especially are significantly different. When the whisker content is 30%, the impact strength of the composite material produced by the twin-screw extruder is up to 10.18 kJ/m², whereas the impact strength of composite produced by a single-screw extruder is 3.3 kJ/m². As a result, the mixing effect of equipment will

Table 4.3 Influence of Different Extrusion Equipment on PP Performances Filled with Calcium Sulfate Whiskers

Equipment	Whisker Content (%)	Tensile Strength (MPa)	Elongation at Break (%)	Bending Strength (MPa)	Impact Strength (KJ/m)	Melt Flow Rate (g/10 min)
Single-screw extruder	30	20.2	9.2	38.1	3.3	4.4
	40	13.5	3.1	29.2	2.1	4.5
Twin-screw extruder	30	26.5	7.8	59.4	10.2	2.1
	40	17.3	3.2	60.2	3.24	1.9

directly affect the mechanical properties of whisker-reinforced composites. Twin-screw extrusion equipment can achieve better mixing and dispersion between whiskers and resins, thereby producing a good enhancement effect of whiskers.

In addition, from the melt flow rate data in Table 4.3 we can see that the melt flow rate of composite material made from single-screw extrusion equipment is significantly higher than the corresponding values from twin-screw extrusion equipment.

4.2.1.2 Milling Blending Method

In the milling blending method, mixing and dispersion are conducted in a double-roller open mill and then molded.

An open mill is a short name for an open plastic mixing mill. It is a roller exposed rubber machine used in a rubber factory to prepare plasticated rubber and rubber compounds, or to conduct heat refining, and it is part of the basic equipment in the rubber industry. An open mill mainly relies on two relative rotary rollers to extrude and shear rubber material and cut the macromolecule chains inside rubber through multiple kneading and associated chemical reactions in the process of kneading, which evenly blends all kinds of components inside rubber and finally achieves the purpose of milling.

The primary working parts in an open mill are two hollow rollers or drilling rollers that rotate inward in different directions. The front roll on the operator's side can be moved back and forth by a manual or electric operation to adjust roller distance to adapt to the operation requirements. The back roller is fixed and cannot be moved back and forth. Two rollers are generally at the same sizes and relatively rotate at different speeds. Raw rubber or plastic material passes into the gap between two rollers with the rotation of rollers, achieving plastication or mixing through strong shearing action. Figure 4.5 is a schematic diagram of an open mill.

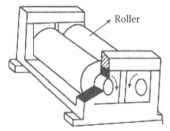

Roller

Figure 4.5   Schematic diagram of an open mill.

An open mill is also used in plastic processing departments and is short for an open plastic mixing mill. In a plastic factory, people generally call it a two-roller machine. An open mill is a kind of mixing plastic equipment applied earlier in a plastic products factory. Its role is to blend and plasticize evenly mixed raw materials and form uniformly mixed and plasticated molten material. When cable materials are produced, an open mill could directly plasticize mixed powder into molten material.

An open mill has a simple structure, can be made easily, and has easy operation and convenient maintenance and disassembly; it is therefore widely used in plastic product enterprises. The disadvantage is a large physical exertion of operators. Under high temperature, the mixing material has to be flipped by hand, and the flip frequency of mixing plastic by hand greatly influences the mixing quality of raw materials.

The specifications of an open mill are represented by diameter × length of the working section of the roller, such as 550 × 1500, and the units are in millimeters (mm).

The representative method of the national standard in China is to add the Chinese phonetic alphabet before the roller diameter number to indicate the use of the machine. Because most front and rear rollers have the same diameters, therefore, the L/D ratio is specified by the national standard, and generally is represented by roller diameter. For example, in XK-400, X represents rubber, K represents an open mill, and 400 represents the diameter of the roller; in SK-400, S represents plastic, K represents an open mill, and 400 is the diameter of the roller; in X (S) K-400, X represents general rubber and plastic,

K represents an open mill, and 400 is the diameter of the roller. For some dedicated open mills with special functions, one more letter is added, for example, in SKP-400, P represents a rubber crushing machine; in XKA-400, A represents a thermal refining machine.

In the case of PVC filled with calcium carbonate whiskers, the operations are as follows. Take 100 parts by weight of PVC as the matrix resin, add 0.2 part by weight of $PbSO_4$, 1 part by weight of plasticizer (DOP), 1 part by weight of acrylate copolymer, 10 parts by weight of chlorinated polyethylene, 0.2 part by weight of paraffin, 3 parts by weight of composite stabilizer, and a calcium carbonate whisker with a mass fraction of 4.2%; mix mix them at room temperature for 5 minutes in a high-speed mixer and then blend on a twin-roller open mill at about 170°C.[8]

4.2.1.3 Internal Mixing and Blending Method

The mixing device of the internal mixing and blending method is an internal mixer, first mixing then molding. An internal mixer is an interval mixing device with high strength, which is developed on the basis of an open mill. It has a series of excellent features that make it better than an open mill: large mixing capacity, short processing time, high efficiency, better ability to overcome dust float, reduced loss of chelating agent, improvement in product quality and working environment, safe and convenient operation, and less labor intensity; it also is beneficial in realizing mechanization and automation. An internal mixer is still a typical and important equipment in plastication and mixing, and is still in constant development and improvement.

An internal mixer has a pair of relatively rotating rotors with a particular shape, and is used for interstitially plasticating and mixing polymer material under airtight conditions with adjustable temperature and pressure. It is made of a mixer chamber, rotors, rotor sealing device, loading and pressing devices, discharging device, transmission device and motor base, and so forth. The working sequence is as follows. Two rotors relatively rotate, which passes the material from the feeding inlet clamp into the roller gap and the material is compressed and sheared by rotors, after which it is divided into two parts, which go back to the top of the roller gap through the gap between chamber walls and rotors, respectively. The material is under shear and friction action when it flows around rotors, which rapidly increases melt temperature and reduces viscosity. At the same time, rotors stir constantly and make the melt material move along the axial of rotors, which results in uniform mixing of the material. Because of the high temperature of the internal mixer during mixing, the shearing action on the melt material is much larger than that of an open mill, which makes the efficiency of an internal mill much higher than that of an open mill. Figure 4.6 is a diagram of an internal mixer.

In the study of filling modification, a torque rheometer is a common piece of small test equipment. Because it is equipped with a corresponding internal mixer chamber, the material shows an opposite reaction on rotors in the process of mixing. The opposite reaction is measured by a force transducer and converted into torque value, and the torque value reflects

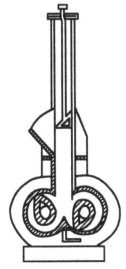

Figure 4.6   Diagram of an internal mixer.

material viscosity. A torque rheometer can be used to study the thermal stability, shear stability, flow, and solidification behavior of thermoplastic polymer materials. Its greatest characteristic is to continually and accurately measure the rheological proper-ties of the system in the machining process similar to the actual production. For example, add a certain amount of PVC to XSS-300 type torque rheometer and preheat to a certain tempera-ture, then add modified magnesium hydroxide whiskers and mix, which leads to even distribution of the whiskers in PVC. When the content of the whisker reaches 30%, the elastic mod-ulus of the composite is 0.24 GP, which is 41.18% higher than the pure PVC.[9]

4.2.2 Solution Dispersion Method

In the solution dispersion method, first matrix resin is dissolved in solvent; then treated inorganic whiskers are added and well stirred until evenly dispersed in solution. Next solvent is removed or polymerized, and a composite material is obtained.

Considering epoxy resin filled with potassium titanate whisker as an example,[10] the specific operation is as follows. Mix bisphe-nol A–glycidyl methacrylate (bis-GMA) and triethylene–glycol dimethacrylate at a mass ratio of 1:1 as a resin matrix; add ini-tiator benzoyl peroxide (BPO) and the polymerization inhibitor butylated hydroxytoluene (BHT, 264) and blend uniformly. Then mix surface-modified potassium titanate whisker into the resin, fill into the mold, and cure in an oven at 120°C for 30 minutes.

The method for bismaleimide resin filled with whiskers is similar to the preceding.[11] The mass ratio of BMI (bisma-leimide) and BA (diallyl bisphenol A) is 1:0.8. First, heat BA to 130°C, then add BMI slowly, prepolymerize for 20 minutes, add the whisker, mix uniformly, pour into a glass mold coated with a release agent, vacuum and deaerate at 120–130°C for 15 minutes, and cure according to the following sequence: 145°C/2 h + 160°C/2 h + 180°C/2 h + 200°C/2 h. Finally, post-treat at 220°C for 6 h.

The characteristics of the solution dispersion method are low temperature, small viscosity of solution, and more uniform dispersion. However, this method is used mostly for the preparation of composite material of thermosetting resin filled with whiskers.

4.2.3 In situ *Polymerization*

In situ polymerization is a method in which inorganic whiskers are evenly dispersed in reactive monomers (or their soluble prepolymers) and then are polymerized using a method identical to bulk polymerization. At the beginning of the reaction, monomers are prepolymerized and then deposited on the surface of whiskers after the prepolymers increasingly aggregate to obtain composite materials filled with whiskers. In this way, inorganic whiskers can be uniformly dispersed in a polymer matrix, and the composite material obtained has good processing performance.

For example, Chen et al.[12] prepared zinc oxide whisker/polyaniline core–shell composite material using *in situ* polymerization. The specific operation is as follows. A certain amount of treated zinc oxide whiskers, aniline, and ethanol are well mixed; then the appropriate amount of 0.01 mol/L HCl is added dropwise with stirring in an ice bath of 0–2°C to initiate polymerization and the mixture is reacted for 10 minutes. The reaction mixture is centrifuged, filtered, and then dried in a vacuum furnace at 50°C for 24 h.

4.3 Performance Analysis of Polymer Matrix Composites Filled with Inorganic Whiskers

4.3.1 Cost Analysis

It is generally believed that because the price of packing materials is much lower than that of polymer resins if calculated based on qualities, the price of composite materials filled with packing materials will decrease significantly. At present some companies still sell plastic films, flat wires, and packing tapes according to their quality (weight on the market). However, because the density of inorganic mineral filler is much higher than that of plastic resin and the usable area or length of filling product in unit weight is less than that of the pure products, sales at the same price will impair the benefit to consumers. Consumers will force down the price of the products after some thought, and therefore calculating economic benefits by weight cost is not scientific.

The cost calculation of filled resin should take the volume cost of the final filled products as an indicator because the price unit of products is number (such as toys, pots, barrels) or usable area (such as thin film). In the case of plastic products, because the density of inorganic mineral filler is much higher than that of the matrix resin, and the mold cavity volume of plastic products prepared by injection molding or compression molding is fixed and should be filled when forming, more materials will be needed to obtain the plastic products with the desired shape and size. That is to say, because packing material has a small volume compared to materials with the same quality (weight), the total volume of filled plastics will be smaller than the volume of the pure matrix resin, and the number of molding products will be lower than the number of the same molding products made of pure resin. Therefore, sometimes it is not worthwhile to add packing material, and this situation should be considered.

Of course, it is another matter if the purpose of using packing material in injection or molding products is not only to reduce the cost of raw materials but also to achieve other aspects.

Although the price of inorganic whiskers is lower than in the past, it is still higher than that of other inorganic mineral packing. The price of relatively cheap calcium sulfate whiskers and calcium carbonate whiskers is hundreds of US dollars per tonne, but the price of calcium carbonate is tens of US dollars per tonne. Thus it can be seen that high cost hampers the development and application of whiskers, and reducing the cost of whiskers will help the development and application of whiskers as fillers.[1,13]

4.3.2 Rheological Behavior

The polymer will begin to melt and become a viscous flow state when the temperature is higher than T_m (melting point) or T_f (viscous flow transition temperature). The mechanical property of a polymer melt is rheology, which is the flow under the action of an external force. The polymer molding is usually processed in a viscous flow state, such as extrusion, injection, compression, rolling, and so forth, whereas most polymer melts are non-Newtonian fluids with high viscosity $(1-10^6 \text{ Pa·S})$. The shear stress and shear rate of polymer melt change greatly in various molding processes, which makes the study of rheological behavior of polymer melts very important.

The flow of polymer melts in extruders, injection molding machines, and pipes with the same cross section is mostly shear flow, which can be represented by the power law equation:

$$\sigma_\tau = \eta \times \dot{\gamma}^n \tag{4.1}$$

where
σ_τ = shear stress (Pa)
γ = shearing rate (s⁻¹)
n = rheology index, also called non-Newtonian factor, representing the degree of deviation of non-Newtonian fluid and Newtonian fluid
η = melt viscosity (Pa·s); viscosity is the index of liquidity of the polymer melt or solution

The current studies of whisker materials in China are focused mostly on the mechanical properties of polymer composite systems, and there is less research on rheology of polymers. Figures 4.7 and 4.8 show the influence of filling content of PP filled with calcium carbonate whiskers and light calcium carbonate on rheological properties, respectively. Table 4.4 shows calculated non-Newtonian factors for fillings with two packing materials.[1,14,15]

From Table 4.4 we can see that the introduction of filler material increases the non-Newtonian property of melts, and

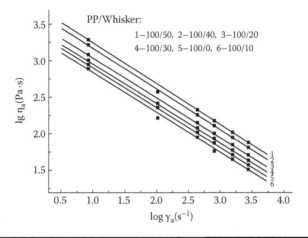

Figure 4.7   Influence of the filling content of calcium carbonate whiskers on rheological properties of polypropylene matrix (180°C).

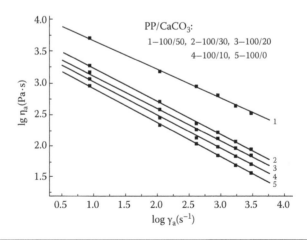

Figure 4.8   Influence of the filling content of light calcium carbonate on rheological properties of polypropylene matrix (180°C).

Table 4.4   Influence of Filling Amount of Different Packings for Polypropylene on Non-Newtonian Factors (*n*)

Filler	Ratio of Matrix to Filler	Non-Newtonian Factor (n)	Correlation Coefficients (R)
CaCO$_3$ Whisker	100/0	0.540	0.997
	100/10	0.539	0.997
	100/20	0.535	0.996
	100/30	0.534	0.996
	100/40	0.552	0.998
	100/50	0.552	0.998
Light CaCO$_3$	100/0	0.540	0.997
	100/10	0.527	0.999
	100/20	0.511	0.999
	100/30	0.520	0.999
	100/50	0.454	0.998

the whisker material has a better influence on the rheological property of the filled composite system than the light calcium carbonate does. With the same filling content, whiskers have a much smaller effect on the liquidity of polymers, which makes the flow characteristic of the composite system deviate much less from the Newtonian fluid.

The addition of fillers importantly influences melt viscosity and is mainly reflected in the following aspects.

1. Different shapes of the packings have different effects on the viscosity of the filling system. For packings with the same *L/D* ratio, flake shape fillers have a greater influence on the viscosity of the filling system than fibrous fillers.
2. The smaller the particle size of the filler, the easier they aggregate together. The filler in the aggregated state has an unfavorable effect on the liquidity of the filling system.

Pretreating the surface of fillers can reduce the surface energy of fillers, which can prevent the aggregation.

3. The dispersion of fillers in the matrix resin has a significant influence on the viscosity of the filling system. If fillers are not well dispersed, they will exist in a certain aggregation state, which adversely affects liquidity.

Einstein studied the influence of the filler concentration on the viscosity of the dispersed particle system, and presented the following equation:

$$\eta = \eta_1(1 + K_g v_2) \tag{4.2}$$

where η is the viscosity without filler; v_2 is the filler concentration; K_g is 2.5 for a hard sphere and normal fluid; for fibrous fillers, K_g increases with an increase of the L/D ratio; the K_g value of the filler with uniaxial orientation is two times the L/D ratio. The preceding is for a low-concentration situation. The equation can be corrected for a dispersed particle system with a high concentration; for example, Simba and Guth presented the following equation:

$$\eta = \eta_1(1 + K_g v_2 + 1.41 \times v_2^2) \tag{4.3}$$

For mixed filling with different states, equation can be properly corrected to characterize the relationship between the melt viscosity of the filling system and the variety, shape, particle size, and concentration of fillers.[1]

Many quick and easy measuring methods are usually used in the industry to measure the viscosity, and melt flow index (MFI) is one of them. MFI is the indicator of the liquidity of melts, defined as the mass of polymer flows through a specific capillary in 10 minutes under a certain pressure and temperature. The greater the MFI, the better is the liquidity and the lower the viscosity. Lower melt viscosity will be good for filler dispersion and processing of composite materials. The MFI can be measured using a standard melt index apparatus.

Figure 4.9 shows the changing curve of the MFI of PP filled with different amounts of whiskers. M-PP represents PP modified with basic magnesium sulfate whiskers; K-PP represents PP treated with potassium titanate whiskers.

From Figure 4.9 we can see that with the increase of whisker content, the MFIs of both systems increase first and then decrease. When the content is 10%, the MFIs of both systems are the largest. The reason is that whiskers have very small sizes and a relatively large L/D ratio; after treatment with a coupling agent, a good interfacial layer forms between whiskers and the matrix resin, and whiskers become arranged in an orderly way in the direction of the melt flow, which reduces the resistance of the melt flow and gives a certain lubrication effect. Furthermore, when the filling contents of whiskers in two systems are both 10%, the MFI of the potassium titanate whisker-filled system is larger than that of the one filled with magnesium salt whiskers. That is to say, the viscosity of the system modified with potassium titanate whiskers is lower.[16]

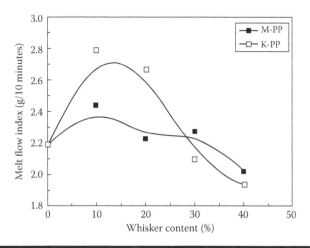

Figure 4.9   Influence of different whiskers on the melt flow index of polypropylene.

4.3.3 Mechanical Performance

The most important performances of polymer materials are their mechanical properties.[1,2] With the addition of whiskers, the original mechanical properties of the matrix resin may change in a way that we want or in an unexpected way.

When materials are affected by external forces without inertial movement, their geometrical shapes and dimensions will change, and this change is called strain or deformation. When materials deform macroscopically, their internal molecules and atoms relatively displace, which brings an additional force against external forces between molecules and atoms. When a balance is reached, the additional internal force is equal to external forces with opposite directions. The internal force per unit area is defined as strain, and its value is equal to that of the external forces. Materials deform in different ways when stressed differently. For the same material, there are three basic types of deformation: simple tension, simple shear, and uniform compression. A material is in simple tension when it is affected by two forces that are perpendicular to the section, equal and opposite in direction, and in the same straight line; a material has a sheer reaction when it is affected by two forces that are parallel to the section, in equal and opposite direction, and at different straight lines. Uniform compression occurs when the material is surrounded by stress p and the volume decreases.

We briefly discuss the following mechanical properties of a polymer matrix composite material filled with whiskers.

4.3.3.1 Strength

Strength is a measurement of the ability of a material to resist damage caused by external forces, which characterizes the mechanical limit of a material and plays an important role in practical applications. Different damaging forces correspond to different strength indexes.

4.3.3.1.1 Tensile Strength

Tensile strength (TS) or ultimate strength is the maximum stress that a material can withstand while being stretched or pulled along an axial direction before failing or breaking under a certain temperature, moisture, and loading rate, and the value is the ratio of the maximum stress (F) to the original cross-sectional area of the initial sample (A_0). Tensile strength is calculated according to the following formula:

$$\sigma = \frac{F}{A_0} \qquad (4.4)$$

When a material undergoes a large deformation, the cross-sectional area also undergoes a larger change. At this time the real cross-sectional area (A) should replace A_0, and the corresponding real stress is called the true stress and is shown as σ'.

$$\sigma' = \frac{F}{A} \qquad (4.5)$$

The corresponding true strain (δ) is as follows:

$$\delta = \int_0^l \frac{dl_i}{l_i} = \ln \frac{l}{l_0} \qquad (4.6)$$

Packing material is the dispersed phase in a filling system and is actually divided in the continuous phase of matrix resin. It is assumed that the gaps among packing particles are completely filled with matrix resin without holes or bubbles, which makes the area of the forcing section of the matrix resin less than that of the material composed of pure resin. Under the action of a tensile force, matrix resin will leave the packing material and produce micro holes that have a different refractive index from that of the surrounding material and therefore the color will be whiter than that of the original material.

4.3.3.1.2 Bending Strength

In a bending test a static bending moment is added to standard samples under certain conditions until the samples break. The maximum stress before breaking is called the bending strength, also known as flexural strength. The elastic modulus under a small deformation is called the bending modulus.

Flexural strength is calculated according to the following formula:

$$\sigma_t = 1.5 \frac{PL_0}{bd^2} \tag{4.7}$$

The bending modulus is

$$E_t = \frac{\Delta PL_0}{4bd^3\delta_0} \tag{4.8}$$

where P is the maximum load; L_0, b, and d represent the length, width, and height of the sample; and ΔP and δ are the load and deflection under a small bending deformation.

Generally speaking, for most packing materials, the bending strength of filling polymers decreases with increasing fraction of the packing material. The declining degree is related not only to the toughness of the matrix resin and the geometry of the filler, but also to the dispersion of the filler in the matrix resin and the processing orientation. A filler with a large L/D ratio or treated with a coupling agent can increase the flexural strength of the polymer.

4.3.3.1.3 Impact Strength

Impact strength is generally defined as the energy absorbed in unit area when a sample is broken under an impact load. It describes the ability of a material to absorb shock and impact energy without breaking. There are many methods to measure impact strength, such as a pendulum, drop method, and high-speed drawing method. The values are different when measured by different methods.

The following is the formula to calculate impact strength:

$$\sigma_1 = \frac{W}{bd} \ (kJ/m^2) \tag{4.9}$$

where W is the consumed power and b and d represent the width and thickness of the material. If the sample has a gap, then b and d stand for the width and thickness of the gap.

The most common impact tester is a pendulum. A pendulum tester is divided into a beam (Charpy) type and a cantilever beam (Izod) type according to the way the sample sets. Samples either can have a gap or be without a gap.

The addition of inorganic mineral particles tends to decrease the impact resistance of filling plastics, which is an important aspect of performance degradation of a material after modification with inorganic particles to achieve a variety of benefits simultaneously. The main reason is that inorganic particles are rigid and therefore cannot deform under an exterior force. They cannot terminate a crack or produce craze to absorb impact energy, and thus will increase the brittleness of filled plastics.

Inorganic whiskers are fibrous, and have high strength and high modulus. Fibrous whiskers can produce a certain deformation under stress and make it easy to relax stress, which helps to eliminate interfacial stress and residual stress and reduce the internal stress of products. The orientation of whiskers can effectively deliver stress, prevent crack propagation, reduce the formation of defects, and improve mechanical strength, thereby achieving enhancement and toughening. For example, when PP is filled with 15% calcium carbonate whiskers, the tensile fracture strength of the $CaCO_3$/PP composite material increases by 35.7%, bending modulus increases by 117%, and impact strength is enhanced by 31.5%.[17] When PVC is filled with 11.5% calcium carbonate whiskers, the impact strength is 12.71 kJ/m^2, which is 254% higher than that of PVC material without whiskers.[8]

It is worth noting that although the improvement in the strength of a polymer system filled with inorganic whiskers is due to the high strength and high modulus of whiskers and a certain L/D ratio, the L/D ratio is far lower than that of glass fiber and carbon fiber. Thus, only when the density of the whisker in the polymer system is greater than a certain minimum value, they can pack closely and transfer strain from whisker to whisker. When the dosage of the whisker is too low, whiskers not only cannot enhance performance, but instead become excess impurity and even a source of a defect and lead to a decrease in the strength of composite materials.[18] Second, with the increasing dosage of packing material, more whiskers will transfer loads, bridge cracks, and deflect cracks. When a composite material is under stress, whiskers locally resist strain in the surrounding matrix, which brings a stronger stress to the whiskers and the whiskers produce a certain deformation under stress. This process relaxes the stress relax, eliminates the concentration of interfacial stress and residual stress, and absorbs impact vibration energy, thereby reducing the stress on substrate materials. At the same time, the extension of the crack in the presence of whiskers is blocked, and therefore cracks are prevented and absorb more energy. In addition, whiskers are pulled out from the matrix because of the consumption of the energy of the external load due to interfacial friction, thereby achieving a toughening effect. Third, whiskers are added to the resin after surface treatment, spread evenly, and play the role of skeleton, thereby reduce the defect, effectively transferring stress and preventing crack extension.[19]

4.3.3.2 Modulus of Elasticity

The modulus of elasticity, often referred to as the modulus, is the stress required on the unit strain. The modulus of elasticity is called tensile modulus (E), shear modulus (G), and compression modulus (B) corresponding to tension, shear, and compression, respectively. Tensile modulus is also known as Young's modulus.

$$E = \frac{\sigma}{\varepsilon} \tag{4.10}$$

$$\sigma_s = \frac{\sigma_s}{\gamma} \tag{4.11}$$

$$B = \frac{P}{\gamma_v} \tag{4.12}$$

The relationship of these three moduli is

$$E = 2G(1 + v) = 3B(1 - 2v) \tag{4.13}$$

where v is Poisson's ratio, defined as the ratio of lateral strain to longitudinal strain in tensile deformation.

Modulus is the characterization of a rigid material. The elasticity modulus of plastic products made from pure resin is low, even for polyesters and polyamides with relatively higher elastic moduli, which is only 2.5%–10% of the metal elasticity modulus. The modulus of elasticity of filled plastics always increases with the addition of filler because the filler modulus is much larger than the modulus of polymers. For example, when 40 phr of magnesium sulfate whiskers are added to PVC, the breaking strength of the composite material increases by 62.43% and the modulus of elasticity increases by 52.38%.[20]

4.3.3.3 Elongation at Break

In the stretching process of external forces that are perpendicular to the section, equal and opposite in direction, the deformation of the material is called tensile strain. When elongation is small, the tensile strain is

$$\varepsilon = \frac{l - l_0}{l_0} = \frac{\Delta l}{l_0} \qquad (4.14)$$

where l_0 is the initial length of the material; l is the length after stretching; and Δl is the absolute elongation.

This definition is widely used in engineering and referred to as relative elongation and abbreviated as elongation.

The elongation of a sample at break is called elongation at break, and defined as the percentage of tensile strain at break:

$$\varepsilon_{max} = \frac{(l_{max} - l_0)}{l_0} \times 100\% \qquad (4.15)$$

The elongation at break under tensile stress is decreased in a filling system because of the existence of packing material, and the main reason is that the majority of packing materials, especially inorganic mineral fillers, are rigid, and most are particles. When a polymer resin is modified by inorganic whiskers, its elongation increases due to the fibrous structure of whiskers.

4.3.3.4 Hardness

Hardness is an indicator of a material surface to resist mechanical stress. Hardness is related to tensile strength and the elastic modulus of a material, and is therefore sometimes used as an approximate estimation of tensile strength and the elastic modulus.

Hardness can be measured by many methods, including dynamic loading and static loading. The former applies an elastic rebound method and impact force to press in the steel ball. The latter uses a hard material with a certain shape as the head and presses the head into the sample by gradual loading.

The static loading method is divided into Brinell, Rockwell, and Shaw hardness tests because of different shapes of the head and different calculation methods.

The hardness of a plastic is different from that of metal and filler, and it is a measurement of the elasticity modulus of the plastic essence; therefore a packing material that can increase the modulus of the filling plastic can also increase the hardness of the material. Because Shaw's hardness test presses sharp needles into plastic material, whether the touching part is plastic substrates or packing material will affect the pressed depth. For filling materials, ball indentation hardness can more accurately reflect the influences of packing material on material hardness.

When potassium titanate whiskers modified by a coupling agent are filled into PE material with ultrahigh molecular weight, the hardness of composite material increases significantly, as shown in Table 4.5.

From Table 4.5 we can see that the hardness of the composite material filled with unmodified whiskers increases with the increase of whisker content. The hardness of the composite materials filled with whiskers treated with a coupling agent increases to different degrees. When the whisker content is less than 10%, the hardness increases quickly; when the whisker content is more than 10%, the hardness increases slowly. The effect when using silane coupling agent (KH-550)

Table 4.5   Influence of Whiskers Content on the Hardness of Composite Material

Whiskers Content (%)		0	5	10	15	20	25	30
Shore hardness	Untreated	62	69	70	71	72	72	72.5
	Treated by KH-550	62	72	74.8	75	76	77	77
	Treated by NDZ-201	62	70	72	73	73	74	74

is superior to the effect of titanate coupling agent (NDZ201). When the filling content of whiskers modified by KH-550 is 30%, the Shaw's hardness of the sample is up to 77, which is 15 points higher than pure polyethylene.[21]

4.3.3.5 Friction Performance

Friction and abrasion are important mechanical properties of polymers, and are very important for rubber tire design. When plastic material is used for a dynamic seal or in a situation requiring high wear resistance, a low friction factor and high abrasion resistance are usually desired. The concentration of the filling material formed through the melting and molding method is different from the surface to center. The filling material will directly affect the friction factor only when it is exposed to the surface directly. Therefore, using packing with a low friction factor to reduce the friction factor of plastic is effective. The surface of the filling plastic products should be ground, which exposes the packing material with a low friction.

Because inorganic whiskers have high strength and a complete structure, the wear resistances of PTFE and PEEK can be improved by adding potassium titanate whiskers, aluminum borate whiskers, or other inorganic whiskers. Feng et al.[22] filled potassium titanate whiskers into PTFE and prepared a composite material whose wear resistance was greatly improved. The loss of wear is only about 10% that of the pure PTFE. Mass formation and extension of cracks are prevented and fatigue wear is delayed.

4.3.4 Other Properties

4.3.4.1 Electrical Property

The majority of polymers are formed by covalent bonds, have high electrical resistivity with both surface resistance and volume resistance greater than 10^{12}, and are electrical insulators with good insulation performance. Thus, the addition of packing material does not affect the insulation of the matrix. More attention is paid to improving the antistatic property of plastic by the filling modification.

Plastics and their products tend to build up static electricity during production, carrying, contact, friction, collision, and so forth. In many cases, generation and release of static electricity can cause serious accidents and affect the normal work of computers. How to prevent the generation of static electricity has been attracting attention over years and a solution is expected. The antistatic performance of plastic material is strictly required in some situations; for example, the surface resistance of various plastic products and materials used in coal mines must be below 3×10^8 Ω·cm. An organic antistatic agent can greatly reduce the surface resistance of plastic material, but as a result of its mobility, the antistatic effect cannot last. A permanent antistatic plastic can be made by adding a conductive filler.

When polygonal needle-like zinc oxide whiskers are added to matrix resin, adjacent needles overlap with each other and form a conductive path, which makes the transmission of charge possible, gives electrical conductivity to material, and also gives the polymer material a certain antistatic property. It can therefore be used as a filler for antistatic plastic.[23] Wu prepared a conductive membrane with good performance by using zinc oxide whisker as packing material and polyvinyl alcohol and poly(methyl methacrylate) as the matrix.[24]

4.3.4.2 Thermal Performance

Sometimes material is used at high temperature, and sometimes we want to improve the heat resistance of some materials. The heat resistance of a material can be characterized by the thermal deformation temperature and Vicat softening point temperature. The thermal deformation temperature can be achieved by applying a certain stress on the material to reach a certain specified deformation. For example, the thermal deformation temperature testing method for a plastic bending load consists of immersing the sample into a liquid whose temperature is increasing at a constant rate and measuring the temperature of the sample when the targeted bending deformation is reached under the static bending load of a beam type. The Vicat softening point temperature of the thermoplastic plastic is measured when the sample is penetrated to a depth of 1 mm by a 1 mm^2 indentor needle under a certain load in a liquid heat-transfer medium with a constant increase in temperature. Both testing methods can evaluate the heat-resisting performance of a plastic material.

Inorganic whiskers have a high melting point and good heat resistance. They are dispersed in a matrix resin in a fibrous shape and play the role of a skeleton. The existence of whiskers also prevents slippage of the macromolecule in the resin, which improves the glass transition temperature and the thermal deformation temperature, as well as the rigidity.

The thermal stability of a material sometimes is represented by the temperature of 5% decomposition and usually can be determined by a thermogravimetric method. Because inorganic fillers have a high melting point, adding inorganic fillers into resin can improve the heat resistance of the polymer. A composite material is prepared by filling calcium carbonate whiskers into PP material. The thermal decomposition temperature (5% decomposition) of the composite material increased from 366.7 to 400.6°C as the whisker content was enhanced.

4.4 Theoretical Analysis of Polymer Matrix Composites Filled with Inorganic Whiskers

4.4.1 Influence of Interfacial Area State on Mechanical Properties

According to the general principle of filling modified polymer matrix composites, the packing material must reach a certain

L/D ratio, and the interfacial combination status of the filler and matrix also plays an important role to ensure that the packing material has enough of an enhancement effect.

At present, research on polymers modified with inorganic whiskers focuses mainly on mechanical performance analysis and further theoretical analysis is relatively rare. In recent years, the finite element method has been widely used in such studies, which can provide a necessary theoretical basis for the microstructure design of material and macro performance improvement.

Guo et al.[25] took resin matrix composites filled with SiC whiskers as an example and analyzed the effect of the interfacial zone state on mechanical properties of composite materials by finite element method. They hold that when whiskers are randomly distributed in a resin matrix with a certain volume fraction, they demonstrate a periodic distribution from the statistical point of view. It is assumed that the SiC whisker-reinforced phase in resin matrix composites is arranged in one direction; without considering the effect of whisker orientation, the analysis model can be simplified as a rotator shape with the large cylindrical part as the polymer matrix and the small cylindrical part as the whisker. Therefore, a three-dimensional problem turns into an axisymmetric problem, and only a quarter part of the body model is taken in the calculation.

Figure 4.10 shows the plane axisymmetric model of a representative volume unit of SiC whisker-filled resin matrix

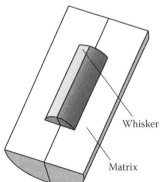

Whisker

Matrix

Figure 4.10   Representative volume unit.

composites simplified using axial symmetry. The model includes the interface between the whisker and the matrix, and can be seen as a layer of interfacial material. The composition of the interface could be measured by the elastic modulus of the interface (E_i). Figure 4.11 is the plane axisymmetric model with Z as the axial direction of the model, X as the radial direction, R as the resin matrix radius, and H as the height; the radius and height of the whisker are r and h, respectively; the height and thickness of the interface between the whisker and resin matrix are h and $0.2r$. The model size is $H = 10h$, $r_i - r = 0.2r$, $h = 10r$, and $H - h = 10r$.

In the stress analysis of composite materials, the parameters are as follows. The elastic modulus of resin matrix $E_m =$ 1.66 GPa, Poisson's ratio $v_m = 0.21$, and yield strength $\sigma_{m\,s} =$ 3.5 MPa; the elastic modulus of the SiC whisker $E_f = 400$ GPa, and Poisson's ratio $v_f = 0.16$; the interfacial layer of the material is isotropic, Poisson's ratio is 0.2, and interfacial elastic modulus is E_i. The elastic-plastic modulus is calculated by using a different interfacial modulus E_i value. The exterior stress σ_0 is set at 0.85 MPa, and the influence of different interfacial bonding states on stress concentration coefficient, interfacial shear stress, and axial stress of whisker end face is simulated. Figure 4.12 is the stress concentration coefficient

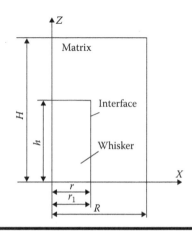

Figure 4.11   Axisymmetric model.

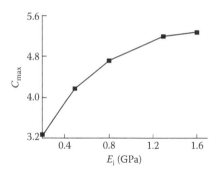

Figure 4.12   Stress concentration coefficient distribution of the whisker end face with different interface elastic moduli.

(C_{max}) distribution of the whisker interface with different elastic moduli.

With the increase of interfacial elastic modulus, the stress concentration of the whisker end face gradually increases, which indicates that a good interfacial bonding condition is an effective way to improve the reinforcing capability of whiskers.

Figure 4.13 shows the distribution curves of axial stress (σ) of composite materials under different interface elastic moduli (E_i).

From Figure 4.13 we can see that when interfacial bonding is good, whiskers undergo large axial stress, whereas when

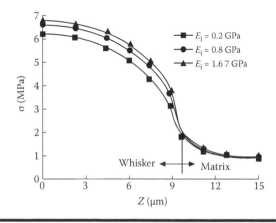

Figure 4.13   Distribution curves of the axial stress of composite materials under different interface elastic moduli (E_i).

interfacial bonding is weak, the axial stress on whiskers significantly decreases and the axial stress on the substrate has no apparent change. The interfacial bonding state has a much greater influence on the axial stress that whiskers can bear than on the resin matrix.

Figure 4.14 shows the distribution curve of the shear stress (τ) of the whisker interface under different interface elastic moduli (E_i). As shown in Figure 4.14, the shear stress on the whisker interface is small when the interface combines weakly, and the shear stress is large when the interface has strong bonding. This is because the whiskers are dispersed phase and resin matrix is continuous phase. A resin matrix cannot pass an external force (σ) to a whisker through interfacial shear stress if the interface combines weakly. Therefore the whisker does not produce any strain, and the stress withstood by the whole composite material is represented by the stress withstood only by the resin matrix. The whiskers originally used for enhancing completely become useless impurities, which damages the integrity of the material and reduces the overall carrying capacity. Thus it can be seen that good interfacial bonding is significant. If not, the composite material will lose its meaning; only when the interface

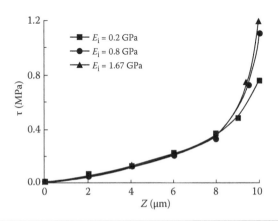

Figure 4.14 **Distribution curve of the shear stress on the whisker interface under different interface elastic moduli (E_i).**

has a certain bonding force can the stress effectively pass to whiskers.

4.4.2 Influence of L/D Ratio on Mechanical Properties

Jia et al.[26] analyzed the influence of SiC whiskers with different L/D ratios on the stress distribution of a SiC resin composite material under a certain SiC whisker volume ratio by using the finite element method on the basis of the axisymmetric model of a resin matrix composite material reinforced by single orientation whiskers, and provided a theoretical basis for the optimal design of a whisker-reinforced composite material.

In the stress analysis of composite materials, the parameters are as follows. The elastic modulus of resin matrix E_m = 1.67 GPa, Poisson's ratio v_m = 0.2, yield strength $\sigma_{m\,s}$ = 3.5 MPa; the elastic modulus of the SiC whisker E_f = 410 GPa, Poisson's ratio v_f = 0.17; the exterior stress σ is set to 0.8 Mpa; and the volume fraction of the whisker in the composite material is 12.5%. The distribution curve of the axial stress of composite material reinforced with whiskers with different L/D ratios is shown in Figure 4.15. In the figure, $Z_1(H)$ stands for the axial

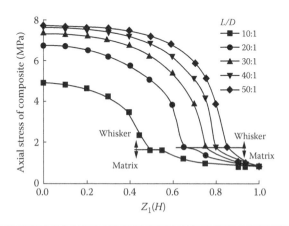

Figure 4.15 **Distribution curve of axial stress of composite material reinforced by whiskers with different *L/D* ratios.**

relative distance from a point on the composite material to the center of the composite material model.

As shown in Figure 4.15, with an increase of the L/D ratio of whiskers, the distribution curve of the axial stress of composite material moves upward, and the stress borne by whiskers and resins increases. At the same time, the stress borne by the resin matrix changes only slightly. The L/D ratio of the whisker has a significantly greater impact on the whisker stress than on the resin matrix.

In addition, in resin matrix composites, stress concentrates on the tip of the whisker with the introduction of the whisker. The maximum stress concentration factor of the SiC whisker is defined as $C_{max} = \sigma_{max}/\sigma_0$. Figure 4.16 shows the stress concentration factor of whiskers with different L/D ratios. The maximum stress concentration factor is C_{max}.

As shown in Figure 4.16, when the L/D ratios of whiskers $L/D \leq 30$, the concentrated stress dramatically increases at the end of the whisker with the increase of the ratio; when $L/D > 30$, a further increase of the ratio has little impact on the maximum concentrated stress.

When a unidirectional composite material is under axial loading, shear stress occurs in the plane parallel to the whisker

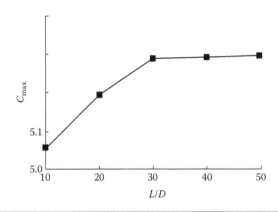

Figure 4.16 Stress concentration factor of whiskers with different *L/D* ratios.

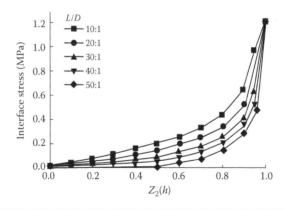

Figure 4.17   Interfacial stress distribution curve of whiskers under different *L/D* ratios.

axle because the elastic moduli of the whisker and matrix do not match. Figure 4.17 represents the shear stress distribution curve of the whisker interface with different *L/D* ratios. $Z_2(h)$ in the figure stands for the relative axial distance from a point in the whisker to the center of the whisker.

As shown in Figure 4.17, with the increase of the *L/D* ratio of whiskers, the interfacial shear stress declines and the curve moves down; but when *L/D* > 30, the ratio has only a slight influence on shear stress.

In the theoretical prediction of axial tensile strength in whisker-reinforced composites, the following mix laws are usually applied to predict tensile strength:

$$\sigma_{cu} = \sigma_{fu}\left[1 - \frac{k_c}{(2k)}\right]v_f + \sigma_{mu}(1 - v_f) \quad k > k_c \qquad (4.16)$$

$$\sigma_{cu} = \sigma_{fu}\left[\frac{k_c}{(2k_c)}\right]v_f + \sigma_{mu}(1 - v_f) \quad k \le k_c \qquad (4.17)$$

where

σ_{cu} = the tensile strength of the composite material

σ_{fu} = the strength of the whisker

k = the ratio of length to diameter of the whisker, $k = L/D$

v_f = the volume percentage content of whisker

σ_{mu} = the stress of matrix as the composite material is damaged

For whisker-reinforced composite material, if the ultimate strength of the whisker is σ_{fu}, the *L/D* ratio of the whisker must be higher than a certain critical value (k_c) to reach σ_{fu}, and k_c is

$$k_c = \frac{\sigma_{fu}}{2\tau} \tag{4.18}$$

where τ is the shear stress of the interface or matrix.

The axial tensile strength at different ratios is shown in Figure 4.18. The tensile strength of the composites gradually increases with an increase in the *L/D* ratio of whiskers.

In fact, the preparation process of composites filled with whiskers often requires two or three procedures. Especially during the surface treatment and blending processes, whiskers often unavoidably break in the process of stirring and mixing, which reduces the *L/D* ratio of whiskers and influences the enhancement and toughening effects of whiskers. Therefore, the performance of the modification can be improved by enhancing the toughness of whiskers and optimizing material preparation technology.

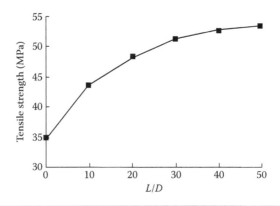

Figure 4.18 **Axial tensile strength of a composite material at different *L/D* ratios.**

4.5 Review of Polymer Matrix Composites Filled with Inorganic Whiskers

In recent years, there has been a breakthrough in the manufacturing cost of potassium titanate whiskers, aluminum borate whiskers, and so forth. Meanwhile, inexpensive and high-quality inorganic whiskers such as calcium carbonate, calcium sulfate, and magnesium are successfully synthesized, which makes inorganic whiskers a new type of filling modification material of polymer materials. This field has become a research hotspot all over the world and reveals broad application and market prospects.

The application of inorganic whiskers in polymer composite materials basically has the following aspects.

4.5.1 Enhancing and Toughening Effect

Inorganic whiskers have both enhancing and toughening effects on polymeric materials. Fibrous inorganic whiskers have almost no interior defects. Because of their high strength and high modulus, they can play the role of a skeleton after being added into thermosetting or thermoplastic resin and form a polymer–fibril reinforced composite material. In addition, when a composite material is under loading stress, the loading stress can be passed through the matrix to whiskers scattered inside. Whiskers can absorb external forces and locally resist strain, thus exerting a toughening effect.

4.5.1.1 Polyolefin Resin

PP is the most studied material in whisker reinforcing and toughening modification research. Zhou et al.[27] studied PP modified with potassium titanate whiskers and found that the notched impact strength increased by 20% over that of pure PP. Lu et al.[28] studied PE composite material modified with magnesium sulfate whiskers, and found that when

10% whiskers were added, the tensile strength of polyethylene increased from 11.5 MPa to 12.2 MPa, which is a 6.1% enhancement, and the tensile strength improved with an increase of whisker content. In a study of PVC composite material filled with magnesium sulfate whiskers, the breaking strength and elastic modulus of the material increased by 62.43% and 52.38% respectively[29] with 40 portion whiskers added, while the notched impact strength of the material increased about 50% with 5 portion whiskers.[30]

It is worth noting the following.

■ When whiskers are treated with different coupling agents, they will have different reinforcing and toughening effects on PP. Zhou et al.[27] filled calcium sulfate whiskers treated with the silane coupling agent KH-550 into PP. The tensile strength of the composite material increased 19% over that of pure PP, but the impact strength decreased.

■ Different processing methods of whiskers have various effects. Liao et al.[31] treated magnesium sulfate whiskers using a titanate coupling agent with dry and wet processing, and the effect of wet processing is apparently better than that of dry processing. When the filler content is 20%, the tensile strength of the material treated by wet processing improved 22.55% over pure PP, and up to 33.7 MPa.

■ Coupling agents have different treatment effects on different whiskers. When PP is filled with potassium titanate whiskers treated with KH-550, the tensile strength and impact strength increase with an increase of filling content, and increase 23.0% and 127.3%, respectively, when the filling content is 35%.[32] When PP is filled with calcium sulfate whiskers treated with KH-550, the tensile strength improves, but the impact strength decreases.[27]

Thus it can be seen that the filling modification effects of inorganic whiskers are related to the variety of whiskers, treatment methods of whiskers, preparation of composite materials filled with whiskers, and many other factors.

4.5.1.2 Epoxy Resin

Whiskers work well in improving the toughness of thermosetting resins because the brittleness of the material decreases and the toughness increases after addition of the whisker. Liu et al.[33] found that the mechanical properties of epoxy resin improve first and then decline with an increase of potassium titanate whisker content. When the mass fraction of whiskers is 5%, the impact strength is greatest, almost double that of the pure epoxy resin. Wang et al.[34] found that when the concentration of aluminum borate whiskers/SiO_2 is 60%, bisphenol F composite epoxy resin has optimum properties. The bending strength is (79.80 ± 2.55) MPa, enhanced 30%; the bending modulus of elasticity is (3616.36 ± 44.00) MPa, increased 1.4 times. Yang et al.[35] also drew the same conclusion. Wang et al.[36] modified polyurethane epoxy resin with calcium sulfate whiskers. When the whisker content was 10%, the shear and peel strength of the resin matrix improved by 5% and 27%, respectively, than the system without whiskers.

4.5.1.3 Polyamide

Cao et al.[37] modified nylon with calcium carbonate whiskers and obtained polymer matrix composites with a high bending elastic modulus, high thermal stability, high surface finish, and good dimensional stability. Li[38] reinforced polyamide (PA) with modified potassium titanate whiskers. While the whisker concentration was 30%, the bending strength of the composite material increased 93% and the tensile strength increased 57%. Shi[39] found that zinc oxide whiskers can significantly improve the impact strength of nylon-6, and a high coupling agent content is beneficial to the improvement of the toughness of the composite.

4.5.1.4 Other Materials

Lei et al.[40] filled magnesium sulfate whiskers into unsaturated polyester resin/ramie fabric composite material. While the filling content of whiskers was 10%, the tensile strength reached the maximum value of 30.16 MPa and improved 5.16% over the composite material without whiskers. Ouyang and Wu[41] added magnesium sulfate whiskers into acrylonitrile butadiene styrene (ABS) resin and prepared a composite material. While whisker content was 50 wt%, the tensile strength was up to 57.72 MPa, increased by 33%. Chen[42] applied calcium metaphosphate whiskers to internal fixation materials of bone fracture. The compressive strength and bending strength nearly doubled compared to pure matrix, and a higher whisker content in materials is beneficial to the proliferation of osteoblasts.

In short, reinforcing and toughening effects of different whiskers vary greatly. Magnesium sulfate whiskers, calcium carbonate whiskers, and calcium sulfate whiskers with high whiteness and low cost will have more applications. Surface treatment of whiskers is also important. The development of new coupling agents and surface treatment agents is also a growing research direction.

4.5.2 Wear-Resistant Effect

4.5.2.1 Polyether Ether Ketone

The most studied anti-wear material modified by whiskers is PEEK. Wang et al.[43] compared the performances of PEEK before and after modification with potassium titanate whiskers. The friction and wear performances of the latter showed obvious improvement over the former. Under 300 N, the wear resistance of the latter increases 2.64 and 2.11 times than the former, respectively. In addition, calcium carbonate whiskers have an excellent anti-friction effect on PEEK composite material. When the whisker content is less than 15%, the wear rate of the material decreases dramatically. The wear rate of the composite material is only 13.2% of the pure PEEK when the whisker content is 15%. Wear rate increases slowly with a further increase of whisker content. When the whisker content is up to 30%, the wear rate is still far lower than that of PEEK.[44]

4.5.2.2 Other Wear-Resistance Materials

Shi et al.[45] also discovered that the wear resistance of PTFE improved about 300 times that of the pure PTFE after adding potassium titanate whiskers. Hu and Liang[46] added aluminum borate whiskers and calcium sulfate whiskers respectively to bismaleimide resin and found that the wearability of both materials improved. When the content of aluminum borate whiskers was 8%, the wear rate of the material declined from 4.39×10^{-6} mm^3/(N·m) to 1.42×10^{-6} mm^3/(N·m). When the content of calcium sulfate whiskers was 10%, the wear rate of the material declined from 2.89×10^{-6} mm^3/(N·m) to 1.28×10^{-6} mm^3/(N·m). Thus it can be seen that the high hardness of inorganic whiskers determines their broad application prospects in the field of polymer wear-resistant materials.

4.5.3 Flame-Retardant Effect

Inorganic whiskers have a high melting point, usually above 1000°C, which just makes up for the deficiency of polymer materials in heat resistance and makes both glass transition temperature and thermal deformation temperature rise and therefore has a flame-retardant effect. The most widely used inorganic whisker is alkali magnesium sulfate whisker [MgSO$_4$·5Mg(OH)$_2$·3H$_2$O], whose crystal water in molecules dehydrates when burning and absorbs a great deal of heat energy to reduce the temperature of the matrix. Furthermore, the water vapor generated not only can dilute the concentration of the reacting gas in the flame zone, but also can absorb smoke and therefore has a flame retardant and smoke suppression effect. Yin et al.[47] confirmed that after adding magnesium sulfate whiskers to a soft PVC matrix, the flame-retardant performance of the composite system improved. When the addition of whiskers is 30 phr and 60 phr, the oxygen indexes of the composites are 26.9% and 28.3%, respectively. The flame-retardant performance has an obvious improvement over that of the pure resin (oxygen index is 25.5%). Furthermore, Han et al.[48] and Liu et al.[49] found that the flame-retardant performance of PP resin is also significantly improved after the addition of modified magnesium sulfate whiskers. Jia et al.[50] added magnesium sulfate whiskers to nylon-6, which improved the flame-retardant property of the composite material.

4.5.4 Conductive and Antistatic Effect

Needle-like zinc oxide whiskers growth in four directions have a three-dimensional extension. When they are dispersed in a polymer matrix, the adjacent needle parts overlap with each other to form a conductive path, which makes the transmission of charge possible, gives the material electrical conductivity, and also brings a certain antistatic property to polymer materials.[23]

He[23] found that zinc oxide whiskers can effectively improve the conductive property of PP resin. When the whisker volume fraction reaches 3%, the resistivity of the material decreases to below 10^9 Ω·cm, which is seven orders of magnitude lower than the resistivity of a pure PP matrix, whose resistivity is 10^{16} Ω·cm and could meet the requirement of a general antistatic material. Wan and Jin[51] put zinc oxide whiskers into an epoxy resin and obtained the same findings. Zhou et al.[52] believed that there are three kinds of conductive mechanisms in the system: network conduction, tunnel effect, and point discharge effect.

4.5.5 Tackifying Effect

Inorganic whiskers with a tiny size can be used in coatings as tackifiers to improve coating viscosity; increase thixotropy;

enhance the surface hardness of paint; and further improve the adhesive force, bonding strength, crack resistance, scratch resistance, scrub resistance, and so forth of the coating.[53] The excellent dispersion effect of inorganic whiskers can also improve the spraying effect of paint, make the construction easier, and improve the production efficiency. In addition, the reasonable collocation with resins also can improve high-temperature resistance and fireproof performance of the coating resin system. Wang et al.[54] studied the influence of calcium carbonate whiskers on silica sol, ethylene–vinyl acetate copolymer emulsion, and styrene acrylic emulsion and found that the addition of whiskers increased the system viscosity by 2 times, 3 times, and 1.3 times, respectively. Zhao[55] studied the influence of calcium carbonate whiskers on the viscosity of epoxy acrylate and urethane acrylate coatings. When the whisker content was 10%, the impact strength of the coatings increased by 40% and 25% over the value of pure matrix resin film, respectively.

4.5.6 Thermal Conductivity

The heat conduction of whiskers relies mainly on lattice vibration, namely phonon vibration; thus thermal conductive performance is excellent, whereas the heat conduction of polymer materials depends mainly on internal vibration between atoms, which makes the thermal conductive performance poor. If whiskers are filled into polymers, they overlap with each other in a matrix and form pathways of heat conduction. Therefore the thermal conductivity of the composite will be greatly increased. Zhou[56] found that low filling content can effectively improve the thermal conductivity performance of epoxy resin. When the whisker content is only 10%, the thermal conductivity of the material increased by 3 times compared with the pure matrix. Li et al.[57] found that the thermal conductivity of PP filled with 30% zinc oxide whiskers increased by 55.9% over pure PP.

4.5.7 Antibacterial Effect

Four-needle zinc oxide (ZnO) whiskers are a new type of antibacterial material, which has high antibacterial efficiency and good color stability compared with other antibacterial agents. It also could increase the strength and toughness of materials. Yamamoto et al.[58] thought that the main antibacterial mechanism of ZnO is that ZnO surface can produce reactive oxygen and water that can permeate the membrane of bacteria, which can destroy the permeability barrier of the bacteria membrane and damage DNA, and therefore inhibit or kill bacteria. Experiments also proved that the ZnO antibacterial agent has relatively high antibacterial activity against *Streptococcus mutans*. The composite resin with strong and long-term antibacterial effects can be obtained by adding ZnO whiskers and the antibacterial rate of the composite resin increases with an increase of the mass fraction of T-ZnOw.[59]

The application of inorganic whiskers has high technical requirements, which involve physics, chemistry, chemical engineering, materials, and other disciplines. Because inorganic whiskers have large polarity, to achieve a good modification effect, surface treatment is required to improve their compatibility with polymer materials. In the process of surface treatment and molding processing, damage to whiskers should be reduced as much as possible to ensure that the whiskers have a sufficient L/D ratio, which is the key to their function. Above all, deeper research is needed to obtain real polymer/whisker composite material.

References

1. Yingjun Liu, Boyuan Liu. *The filling and modification of plastic.* Beijing: China Light Industry Press, 1998.

2. Yan Xia. *The introduction of polymer science.* Beijing: Science Press, 2000.
3. Yanming Dong, Hailiang Zhang. *The introduction of polymer science.* Beijing: Science Press, 2010.
4. Yulong Li. *Polymer additives.* Beijing: Chemical Industry Press, 2008.
5. Zhihua Wu, Qi Yang. *Molding technology of polymer material.* Chengdu: Sichuan University, 2010.
6. Jian Zhou, Jiqin Tang. Study on PP/calcium sulfate whisker composite. *Engineering Plastics Application,* 36(11):19–22, 2008.
7. Tiejun Ge, Hong Yang yi, Yuexin Han. Study on the performance of the calcium sulfate matrix compositing reinforced polypropylene. *Plastic Science & Technology,* (1):16–19, 1997.
8. Litao Li. *Study and application of reinforced and wear-resistant whiskers composites materials.* Wuxi: Jiangnan University, 2008.
9. Yuzhi Jiang, Yuexin Han, Wanzhong Yin. Effect of magnesium hydroxide whiskers on PVC's material mechanical properties. *Metal Mine,* 14(4):172–175, 2009.
10. Xiuyin Zhang, Aifeng Tian, Yirong Liu et al. Potassium titanate whisker modified by silane coupling agent to enhance the bending strength of dental composite resin. *Chinese Journal of Prosthodontics,* 10(2):101–106, 2009.
11. Xiaolan Hu, Zhong Lin, Moufa Yu et al. Tribological properties of bismaleimide resin and its composites. *Journal of Material Engineering* (S). *SAMPE China* 180–182, 2008.
12. Xiaolang Chen, Lv Wangchun, Wei Wang et al. Structure and morphology of pani/t-znow composite. *China Plastics Industry,* 36(7):50–52, 2008.
13. Jianping Dong, Xianlan Ji, Yanwu Lei. Economic evaluation of energy-saving and reduce consumption in reconstruction project. *Chemical Industry,* 25(9):32–35, 2007.
14. Jian Jiao, Weiyuan Lei. *The structure and properties of polymer.* Beijing: Chemical Industry Press, 2003.
15. Ning Liu, Youbing Fan, Dangqing He et al. The effect of rheological properties of polymer/aragonte whisker composites. *Insulating Materials,* 36(1):5–8, 2003.
16. Yukun Wei, Haojiang Wang, Chun Pang et al. Study on properties of polypropylene modified by whiskers. *Synthetic Materials Aging and Application,* 35(1):12–16, 2006.

17. Mingshan Yang, Chenghan He. Preparation of calcium carbonate whiskers and modification of PP with it. *Modern Plastics Processing and Applications*, 20(6):21–24, 2008.
18. Wu Li. *Inorganic whisker.* Beijing: Chemical Industry Press, 2005.
19. Rong Wang, Wenyun Zhang, Anqi Jia et al. Effects of different kinds and usage amounts of whisker fillers on the mechanical properties of resin composite. *Journal of Oral Science Research*, 23(4):365–367, 2007.
20. Yuzhi Jiang, Yanbo Li. Study on magnesium oxysulfate whiskers reinforced PVC. *Transactions of Shenyang Ligong University*, 28(1):1–5, 2009.
21. Sujuan Ye, Qing Fan, Quan Yu. Properties of the PTW filled UHMWPE composites. *Plastics*, 37(6):5–8, 2008.
22. Xin Feng, Donghui Chen, Xiaohong Jiang. Study on friction and wear properties of potassium titanate whiskers-reinforced PTFE composites. *Polymer Materials Science and Engineering*, 20(5):129–132, 2004.
23. Shuhua He, Jianping Zhou, Wanli Fu. Study on electroconductivity of tetragonal zinc oxide whisker (T-ZnOW)/polypropylene composite. *Fine and Specialty Chemicals*, 15(11):29–32, 2007.
24. Huawu Wu, Honghua Huang. Preparation of conductive polymer films containing tetrapod-like ZnO whiskers. *Journal of Functional Materials*, 30(4):389–390, 1999.
25. Weiwei Guo, Ailing Wang, Limei Zhu et al. The effect of interface combination on the mechanical behavior of whisker-reinforced resin matrix composite. *Modern Manufacturing Engineering*, 37(5):124–126, 2009.
26. Haitao Jia, Yongzhen Wang, Ailing Wang. Numerical simulation to effect of whisker aspect ratio on the mechanical behavior of composites. *Modern Manufacturing Engineering*, 35(3):72–74, 2007.
27. Jian Zhou, Yuqi Wang, Haibin Meng. The performance of the potassium titanate whisker modified polypropylene. *Engineering Plastics Application*, 33(11):21–24, 2005.
28. Hongdian Lu, Xiaoling Yu, Jianguo Shao et al. Study on the low density polyethylene/magnesium hydroxide sulfate hydrate whisker composites. *New Chemical Materials*, 35(10):65–67, 2007.
29. Yuzhi Jiang, Yanbo Li. Study on magnesium oxysulfate whiskers reinforced PVC. *Transactions of Shenyang Ligong University*, 28(1):1–5, 2009.

30. Kaizhou Zhang, Shaopeng Yuan, Li He et al. Study on properties of PVC/MSW composites. *China Plastics Industry*, 37(6):18–20, 2009.
31. Mingyi Liao, Xueli Wei. Study on PP reinforced with magnesium hydroxide sulfate hydrate. *Engineering Plastics Application*, 28(1):12–14, 2000.
32. Ning Yang, Dayong Gui, Cheng Qian. Study on potassium titanate whisker reinforced polypropylene. *China Plastics*, 18(1): 71–74, 2004.
33. Lin Liu, Lei Zhao, Lin Li. Application of potassium titanate whiskers in epoxy resin. *Engineering Plastics Application*, 36(3):13–16, 2008.
34. Rong Wang, Wenyun Zhang, Anqi Jia et al. Effects of different kinds and usage amounts of whisker fillers on the mechanical properties of resin composite. *Journal of Oral Science Research*, 23(4):365–367, 2007.
35. Jieying Yang, Guozheng Liang, Yusheng Tang et al. Study on cyana resin/glass fiber composites reinforced by aluminium borate whisker. *Acta Aeronautica et Astronautica Sinica*, 27(2):331–335, 2006.
36. Debo Wang, Jiping Yang, Pengcheng Huang. Adhesive properties of calcium sulfate whisker-modified polyurethane-epoxy resins. *Acta Materiae Compositae Sinica*, 25(4):1–6, 2008.
37. Youming Cao. The preparations and applications of calcium carbonate whisker. *New Chemical Materials*, 26(10):23–25, 1998.
38. Zizhi Li. Study on potassium titanate whiskers reinforced polyamide. *Journal Harbin University of Science & Technology*, 13(4):94–96, 2008.
39. Jing Shi. *Study on tetra-needle-shaped zinc oxide whisker modifying nylon 6 and high density polyethylene*. Chengdu: Southwest Jiaotong University, 2008.
40. Wen Lei, Tao Yang, Chao Ren. Mechanical properties and thermal performance of UP resin/ramie cloth/basic magnesium sulfated whisker composites. *China Plastics*, 20(12):23–27, 2006.
41. Yexian Ouyang, Qingsong Wu. Study on mechanical properties of M-HOS refined ABS composites. *Journal of Wuhan Engineering Institute*, 19(3):1–5, 2007.
42. Lin Chen. *Preparation and properties of β-calcium metaphosphate whiskers/poly-L- laetide composites for internal fixation of bone fractures*. Chengdu: Sichuan University, 2007.

43. Huaiyuan Wang, Xin Feng, Yijun Shi et al. Tribological behavior of PEEK composites filled with various surface modified whiskers. *Journal of University of Science and Technology Beijing*, 29(2):182–185, 2007.
44. Youxi Lin, Chenghui Gao. Effect of CaCO₃ whisker treated with SPEEK on friction and wear of filled PEEK composites. *Lubrication Engineering*, 32(11):65–68, 2007.
45. Yijun Shi, Ming He, Xiaoli Gu et al. A study on the friction and wear mechanism of potassium titanate whiskers filled PTFE composites. *Lubrication Engineering*, 34(1):59–62, 2009.
46. Xiaolan Hu, Guozheng Liang. Friction and wear properties of aluminum borate whiskers modified bismaleimide resin. *Acta Materiae Compositae Sinica*, 21(6):21–26, 2004.
47. Jianjun Yin, Jingjing Li, Yunhua Song et al. Effect of basic magnesium sulfate whisker on properties of flexible polyvinyl chloride. *Journal of Lanzhou University of Technology*, 35(3):21–24, 2009.
48. Yuexin Han, Xu Chen, Wanzhong Yin. The research of basic magnesium sulfate whisker filed up polypropylene. *Non-Ferrous Mining and Metallurgy*, 23(5):42–45, 2007.
49. Qingfeng Liu, Zhou Wang, Wenyu Shang et al. Preparation of calcium carbonate whisker. *Inorganic Chemicals Industry*, 32(2):11–12, 2000.
50. Huawei Jia, ZaiJiang Ding. Study of nylon 6 filled with magnesium sulfate whisker. *China Plastics Industry*, 35(S1):141–143, 2007.
51. Cuifeng Wan, Shengming Jin. Surface modification of T-ZnO whisker and its application in antistatic epoxy resin. *Materials Review*, 21(S1):375 –377, 2007.
52. Zuowan Zhou, Shikai Liu, Lixia Gu. Studies on the conductive property of polymer/ZnOw composites. *Journal of Functional Materials*, 32(5):492–495, 2001.
53. Lixia Li, Yuexin Han, Wanzhong Yin et al. The application and preparation technology of calcium carbonate whisker. *Non-Ferrous Mining and Metallurgy*, 21(S1):73–74, 2005.
54. Huili Wang, Juanjuan Yang, Bin Liu et al. Application of CaCO₃ whisker in the coating. *Paint & Coatings Industry*, 34(4):52–54, 2004.
55. Ziqian Zhao. *Preparation and application of functional UV-curing coating with CaCO₃ whisker.* Wuxi: Jiangnan University, 2007.

56. Liu Zhou. *Preparation and properties of insulating thermal conductive epoxy resin based composites.* Wuhan: Wuhan University of Technology, 2008.

57. Guangji Li, Hui Feng, Qiyong Tong. Study on the preparation and properties of thermally conductive and electric-insulating T-ZnOw/PP composites. *Materials Research and Application,* 2(4):511–516, 2008.

58. Osamu Yamamoto, Miyako Komatsu, Jun Sawai et al. Effect of lattice constant of zinc oxide on antibacterial characteristics. *Journal of Materials Science: Materials in Medicine,* 15(8):847–851, 2004.

59. Lina Niu, Jihua Chen, Ming Fang et al. Effects of three different zinc oxide incorporation on the antibacterial activity against *Streptococcus mutans* of composite resin. *West China Journal of Stomatology,* 44(4):240–242, 2009.

Chapter 5

Surface Modification and Characterization of Calcium Carbonate Whiskers

Calcium carbonate is an important inorganic filler that is easily available and has a low cost. It has been widely used in household plastics, rubber, the paint and paper industry, and has become an important polymer material filler with a huge potential development. Calcium carbonate can be divided into ground calcium carbonate and light calcium carbonate based on its source. Light calcium carbonate for industrial applications currently includes mainly three kinds: calcite, aragonite, and vaterite.[1] Common calcite generally has some defects, such as holes, impurities, incomplete grain boundary and structure, and so on. Calcium carbonate whiskers with almost no structural defects have the structure of aragonite and substantially overcome these drawbacks.[2] Similar to other whiskers, calcium carbonate whiskers have a slender and integrated structure, with the advantages of high strength, high modulus, high whiteness, high filling capacity, good thermostability, thermal resistance, and insulation. The diameter of calcium carbonate whiskers is 0.5–1.0 µm, and the length is 20–30 µm. Because there is no crystal water of crystallization at the molding temperature, the products will not form crazes due to crystal water evaporation in material. Meanwhile, because it has a special shape and the linear expansion coefficient is close to that of the plastic, the whiskers have good compatibility with the matrix resin and therefore can improve the processing performance of the products, such as the mechanical properties, surface smoothness, and abrasion resistance.[3–8]

Calcium carbonate whiskers are produced with calcite, marble, or limestone as raw materials, and the process is simple. Compared with other whiskers obtained from more complex and high-temperature processes, calcium carbonate whiskers have lower cost and more significant economic value, leading to their large-scale use for industrial and consumer products.[9] In addition, calcium carbonate whiskers are nontoxic and biodegradable in an acidic environment; thus environmental pollution and any unfavorable effects on humans are low as well.[10,11]

The infrared spectrum of calcium carbonate whiskers is shown in Figure 5.1; thermal analysis is shown in Figure 5.2; x-ray diffraction is shown in Figure 5.3; a scanning electron

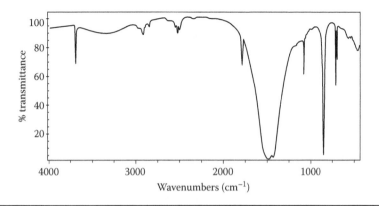

Figure 5.1 **Infrared spectra of calcium carbonate whiskers.**

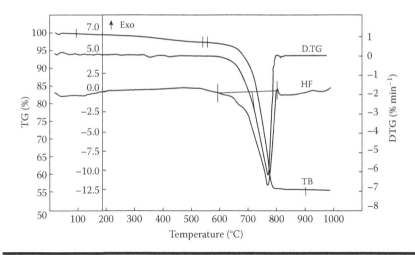

Figure 5.2   Diagram of thermal analysis of calcium carbonate whiskers. (From Wu Li. *Inorganic whisker.* Beijing: Chemical Industry Press, 2005.)

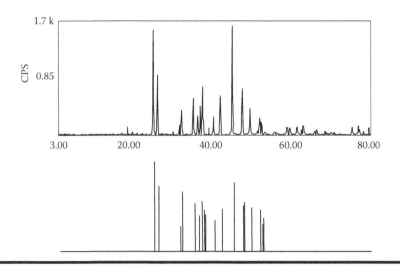

Figure 5.3   X-ray diffraction pattern of calcium carbonate whiskers. (From Wu Li. *Inorganic whisker.* Beijing: Chemical Industry Press, 2005.)

micrograph (SEM) is shown in Figure 5.4; and an energy spectrum is shown in Figure 5.5.

Like other inorganic whiskers, calcium carbonate whiskers have high polarity, and therefore surface modification is required for filling polymer materials. The common surface

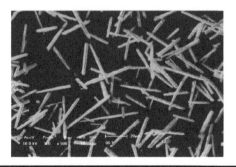

Figure 5.4   SEM photos of calcium carbonate whiskers.

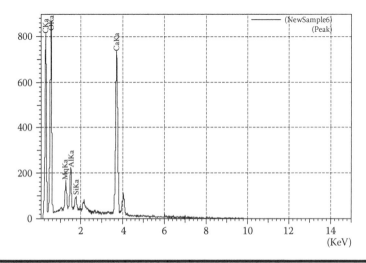

Figure 5.5   Energy spectrum of calcium carbonate whiskers.

modification agents include a higher fatty acid and its salt, titanate coupling agents, silane coupling agents, and so on.

5.1 Modification of Calcium Carbonate Whiskers with Stearic Acid

5.1.1 Modification Method—Wet Processing

Weigh the stearic acid and put it into a beaker. Next add 100 mL of acetone and heat it to dissolve the stearic acid. Shake at about 50°C in a flow thermostatic oscillator, and then add calcium carbonate whiskers while shaking. After all the acetone has volatilized, allow the mixture to stand at room temperature for 8 hours and then place in an oven to dry.

5.1.2 Characterization of Calcium Carbonate Whiskers after Modification

5.1.2.1 Activation Index

The activation index of treated calcium carbonate whiskers is measured. The results before and after modification of calcium carbonate whiskers are shown in Figure 5.6.

Figure 5.6 shows that with the increasing mass fraction of stearic acid, the activation index of whiskers increases at first and then decreases. The mass fraction of the highest activation index is 4.0%; after that, with an increase of the mass fraction of stearic acid, the activation index decreases, which indicates that the activation index reaches the highest value when the whisker surface is completely coated by stearic acid. More stearic acid will cause excessive accumulation and then reduce the activation index.

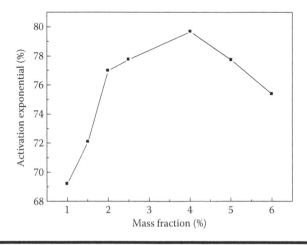

Figure 5.6 **Relationship between the activation index and the mass fraction of modification agent.**

5.1.2.2 Infrared Spectrum Characterization

The infrared spectrum is used for structure characterization; the infrared spectrum of calcium carbonate whiskers before and after treatment with stearic acid is shown in Figure 5.7.

As shown in Figure 5.7, compared with the spectrum of the unmodified calcium carbonate whiskers, the spectrum of the processed whiskers shows a significant increase in the intensity of the absorption peaks at the wave number of 2917 cm⁻¹ and 2849 cm⁻¹. The absorption peak of 2850 cm⁻¹ is the key feature of the symmetric stretching vibration of C–H bonds in CH_3. The absorption peak at 2918 cm⁻¹ shows the asymmetric stretching vibration of C–H, indicating stearic acid molecules with a long fatty acid carbon chain are adsorbed on the surface of calcium carbonate whiskers. However, an ester carbonyl absorption peak at 1700–1800 cm⁻¹ does not appear, indicating that the subpolar ends of stearate molecules are adsorbed on the surface of whiskers by covalent bonding.

Infrared spectra of calcium carbonate whiskers after modification by stearic acid with different concentrations are shown in Figure 5.8. Figure 5.8 shows that with an increase of mass

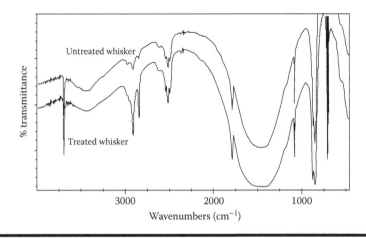

Figure 5.7   Infrared spectra of calcium carbonate whiskers before and after modification.

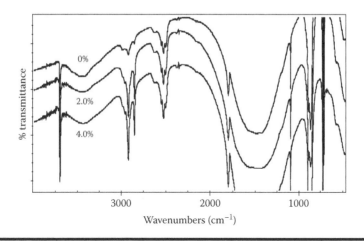

Figure 5.8   Infrared spectra of calcium carbonate whiskers after modification with stearic acid at different concentrations.

fractions of stearic acid, the intensity of the absorption peaks at 2917 cm^{-1} and 2849 cm^{-1} increases significantly.

This modification method has an obvious shortcoming in that excess stearic acid remains on the whisker's surface if not removed. When filling into the polymer, the excess acid molecules will affect the property of materials. We have attempted to fill whiskers modified like this into polypropylene and found that the tensile strength of the composite material is actually lower than that of the pure polypropylene.

5.2 Sodium Stearate Modified Calcium Carbonate Whiskers

5.2.1 Modification Method—Wet Processing

A certain amount of stearic acid is poured into 10 mL of absolute ethanol and heated until it completely dissolves. Sodium hydroxide solution is added dropwise to maintain the pH between 8 and 9. Afterwards, calcium carbonate whiskers are added and stirred for a certain time. Then the sample is collected via filtration, washed and dried, after which modified whiskers are obtained.

5.2.2 Characterization of Calcium Carbonate Whiskers after Modification

5.2.2.1 Effect of Reaction Conditions on the Activation Index of the Whiskers

5.2.2.1.1 Concentration of the Surface Modification Agent

Considering the concentration of the surface modifier is the most important factor affecting the surface properties of whiskers. The experiment is designed at 60°C for 40 minutes. The influence of the dosage of surface modifier to modification effect is investigated, as shown in Figure 5.9.

Figure 5.9 shows that the activation index is almost zero for unmodified calcium carbonate whiskers, but increases significantly after surface modification. Thus it can be seen that sodium stearate is effective in modifying calcium carbonate whiskers. When the concentration of the modifier is less than 4.5%, the activation index of the whiskers gradually increases.

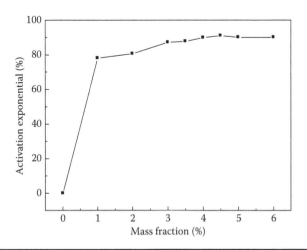

Figure 5.9 Effect of modifier concentration on the activation index of the whiskers.

When the concentration of the modifier increases to 4.5%, the activation index of the whiskers reaches a maximum of 91.0%. When the concentration is greater than 4.5%, the activation index scarcely changes. Overall, the activation index is hardly increased when the concentration of the surface modification agent reaches 3.0%.

5.2.2.1.2 The Effect of Modification Time

The effect of different processing time for a modification effect of calcium carbonate whiskers is shown in Table 5.1.

As shown in Table 5.1, different modification times have different effects on whiskers. The activation index of calcium carbonate whiskers approach a maximum of 93.4% when the processing time is 50 minutes. With an increase of processing time, more modifier molecules are adsorbed on the surface of the whiskers, and the effect of the modification increases gradually. When the processing time is up to 50 minutes, the modification agent can make sufficient contact with the whiskers and fully coat their surface to the maximum activation index. When the modification time extends further, the already adsorbed modification agent on the material surface desorbs, and the modification effect deteriorates.

5.2.2.1.3 Modification Effect of Temperature

The effect of different modification temperature on whiskers modification is shown in Table 5.2.

As can be seen from Table 5.2, the modification temperature has little effect on the activation index of calcium carbonate whiskers, and all the activation indexes are greater than 90%. Taking into account the temperature of the esterification

Table 5.1   Relationship between Different Processing Times and the Activation Index

Time (minutes)	20	30	40	50	60
Activation index (%)	86.7	86.7	91.0	93.4	88.7

Table 5.2 Relationship between Different Processing Temperatures and the Activation Index

Temperature (°C)	30	40	50	60	70	80	90
Activation index (%)	94.6	96.5	96.1	96.0	95.7	95.4	95.8

reaction and the ease of operation, we suggest a modification temperature of 40°C–50°C, which will favor the formation of chemical bonds on the surface of calcium carbonate whiskers.

5.2.2.2 Infrared Spectroscopy

Infrared spectra of unmodified and modified calcium carbonate whiskers are measured and shown in Figure 5.10.

By analyzing the spectra of calcium carbonate before and after modification from Figure 5.10 we can see that the intensity of absorption peaks at 2850 cm^{-1} and 2918 cm^{-1} increases significantly. The absorption peak at 2850 cm^{-1} is the characterization of the symmetric stretching vibration of the C–H bond in CH$_3$. The absorption peak at 2918 cm^{-1} is characteristic of the asymmetrical stretching vibration of the C–H bond in CH$_3$. Those two peaks indicate that sodium stearate molecules with long fatty carbon chains have been adsorbed on the

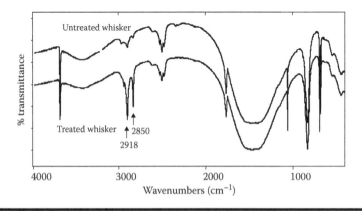

Figure 5.10 Infrared spectra of calcium carbonate whiskers before and after modification.

surface of the whiskers. Furthermore, there are no hydroxyl absorption peaks from 3100 cm^{-1} to 3500 cm^{-1} nor ester carbonyl absorption peaks from 1720 cm^{-1} to 1750 cm^{-1} in the modified spectra, indicating that no chemical adsorption between sodium stearate and calcium carbonate whiskers occurs, only physical adsorption. According to the surface characteristics of calcium carbonate whiskers, the calcium carbonate whiskers with high surface energy can easily absorb water molecules. The carboxyl groups of stearic acid molecules are adsorbed on the surface of whiskers with an orientation according to the physical intermolecular bonding forces. The long nonpolar aliphatic chains stretch out, which increases the hydrophobicity of the whisker surfaces and significantly improves the activation index of the whiskers.

To characterize further the mode of action of sodium stearate molecules and the surface of the whiskers, we apply the same approach to the aluminum borate whiskers. Infrared spectra of aluminum borate whiskers before and after modification are shown in Figure 5.11.

As shown in Figure 5.11, absorption peaks of 2918 cm^{-1} and 2851 cm^{-1} on the spectra of the modified aluminum borate whiskers significantly increase. The absorption peak at 2850 cm^{-1} is characteristic of the C–H bond symmetric stretching

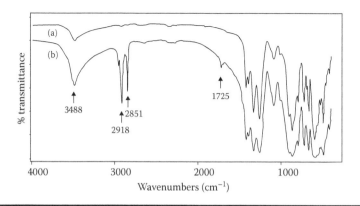

Figure 5.11   Infrared spectra of aluminum borate whiskers before and after modification. (a) Before. (b) After.

vibration in CH_3. The absorption peak at 2918 cm^{-1} is characteristic of the C–H bond asymmetrical stretching vibration in CH_3. The ester carbonyl absorption peak of 1725 cm^{-1} suggests that ester bonds are formed from the hydroxyl on the surface of stearic acid and sodium aluminum borate whiskers. There is both physical adsorption and chemical absorption between sodium stearate and aluminum borate whiskers.

Notably, there is an obvious hydroxyl absorption peak at 3488 cm^{-1} in Figure 5.11, indicating that both the hydroxyl and water molecules are absorbed on the surface of the aluminum borate whiskers. Compared with the spectrum of the unmodified calcium carbonate whiskers in Figure 5.11, presumably more water molecules are adsorbed on the surface of the calcium carbonate whiskers.

There are many studies on calcium carbonate whisker surface modification. Some scholars, for example, Li et al.,[12] believe that the calcium carbonate whisker surface can form chemical bonds with the molecules of stearic acid. Infrared spectra are shown in Figure 5.12.

In addition, there are many types of modifiers applied to calcium carbonate surface modification. We need to

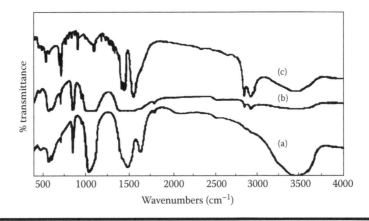

Figure 5.12 **Infrared spectra of calcium carbonate whiskers before and after modification. (a) Calcium carbonate whiskers before modification. (b) Calcium carbonate whiskers after modification. (c) Sodium stearate.**

comprehensively consider the specific surface properties of materials, the use of modified products, quality requirements, modification processes, and cost factors when we choose which one to use. The modification results of different surface modification agents for calcium carbonate whiskers are shown in Table 5.3.

As seen from Table 5.3, the products modified by tita-nate NDZ-101, sodium stearate, stearic acid, and Zinc stearate acid zinc exert the highest activation index; nevertheless, the modification effects of other modifiers are relatively poor. The results of contact angle also support this point. Among these four modifiers, sodium stearate modified products hold the maximum contact angle and have lower costs.[12]

The advantages of the treatment of calcium carbonate whis-kers by wet processing include even dispersion and good effect. The disadvantages are that wet processing is generally done in solution, and therefore a dispersion medium is needed. It also

Table 5.3 Effects of Different Modifiers on the Modification Effect of Calcium Carbonate Whiskers

Surface Modification Agent	Activation Index (%)	Contact Angle (°)
Stearic acid	96.4	118.28
Sodium stearate	99.9	136.79
Zinc stearate	99.9	136.61
Sodium dodecyl sulfate	8.6	–
Sodium dodecyl benzene sulfonate	8.0	–
Titanate NDZ-101	99.9	51.54
Titanate NDZ-401	32.4	–
Aluminum acetate	66.9	79.41
Silane KH-570	12.8	–
Polyacrylamide	0	–

requires filtration and drying processes; the processing method is relatively cumbersome and generally used in the case of a small amount of whiskers, with more used for experimental studies. Furthermore, when modified whiskers are filled into a resin matrix to prepare composite material, because of the high preparation temperature, thermal stability has to be taken into account when choosing a surface modification agent.

If the matrix resin contains processing molding additives, such as a plasticizer, stabilizer, lubricant, and so forth, the different compatibilities among these additives, between additives and resin, and between the filler and additives, make the interfacial area of the filler and the matrix and the interfacial distribution complicated, and the performance of the composite material will be influenced by many factors.

5.3 Modification of Calcium Carbonate Whiskers by a Titanate Coupling Agent

Dry processing is applied more often in industrial production. We adopt dry processing to modify calcium carbonate whiskers and investigate the influence of a titanate coupling agent on the surface properties of the whiskers.

5.3.1 Modification Methods—Dry Processing

The calcium carbonate whiskers are previously dried under 120°C for 8 hours. The coupling agent is diluted with a certain amount of white oil. A certain quantity of dry calcium carbonate whiskers and the coupling agent are put into the high-speed mixer and mixed until uniform. Remove the mixture and add it to the twin-screw extruder, squeeze out, place the extrudate in a flat vulcanizing machine to press molding, cool to room temperature, and then release. Cut out standard specimens through a universal cutting machine and measure the mechanical properties.

5.3.2 *Mechanical Performance Analysis*

NDZ-101 and NDZ-102 titanate coupling agents are selected to treat the calcium carbonate whiskers. The tensile properties of polypropylene composites filled by modified calcium carbonate whiskers are shown in Figures 5.13 and 5.14, respectively.

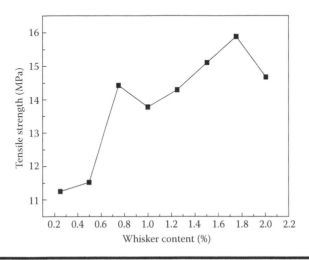

Figure 5.13   Effect of NDZ-101 content on the tensile properties of the composite material.

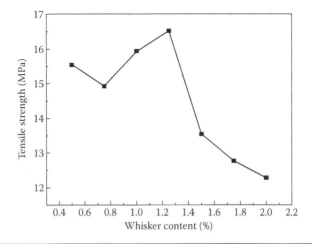

Figure 5.14   Effect of NDZ-102 content on the tensile properties of the composite material.

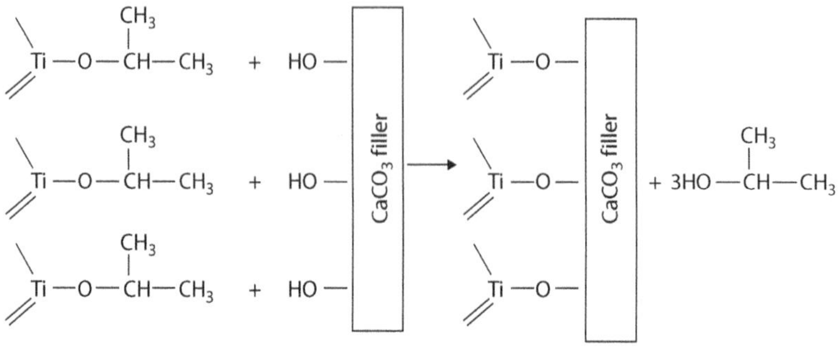

$$\begin{array}{l}
\text{Ti}-O-\overset{\overset{\displaystyle CH_3}{|}}{CH}-CH_3 \quad + \quad HO- \\[4pt]
\text{Ti}-O-\overset{\overset{\displaystyle CH_3}{|}}{CH}-CH_3 \quad + \quad HO- \quad\Big|_{CaCO_3\ filler} \longrightarrow \quad \text{Ti}-O- \quad\Big|_{CaCO_3\ filler} \quad + 3HO-\overset{\overset{\displaystyle CH_3}{|}}{CH}-CH_3 \\[4pt]
\text{Ti}-O-\overset{\overset{\displaystyle CH_3}{|}}{CH}-CH_3 \quad + \quad HO-
\end{array}$$

Figure 5.15   Interaction between calcium carbonate filler and titanate coupling agent.

Figure 5.13 shows that the tensile strength of the composite material increases with an increase of NDZ-101 coupling agent. When the coupling agent content is 1.75%, the tensile strength reaches its maximum value of 15.9 MPa. As we can see from Figure 5.14, with an increasing amount of NDZ-102 coupling agent, the tensile strength of the composites shows a decreasing trend after the first increase. When the content of the coupling agent is 1.25%, tensile strength is the maximum, about 16.5 MPa. By comparing the effects of the two coupling agents, we can see that NDZ-102 is slightly better than the coupling agent NDZ-101 in terms of the modification effect.

The chemical name NDZ-101 is isopropyl dioleateacyloxy (acyloxy phosphatedioctyl) titanate. The chemical name of NDZ-102 is isopropyl tri (dioctyl phosphate acyloxy) titanate. Both are monoalkoxy titanates. The hydrolysis reaction takes place once the coupling agent meets the water molecules on the surface of calcium carbonate whiskers, and then the coupling agent is conjugated to the surface of the whiskers. The effect is shown in Figure 5.15.[13]

5.3.3 SEM Analysis

In the matrix material, the distribution of calcium carbonate whiskers and interfacial compatibility of whiskers and the matrix material greatly impact the properties of composite materials.

The tensile fracture of composite materials is usually observed using SEM. NDZ-101 and NDZ-102 are the coupling agents to modify calcium carbonate whiskers, with 20% filling. The SEM photographs of tensile fractures are shown in Figure 5.16.

As shown in Figure 5.16a, the composites that contain the NDZ-101 coupling agent modified calcium carbonate whiskers present significant brittle fracture, a small number of whiskers are pulled out, and most whiskers show a significant cross-sectional fracture. Figure 5.16b shows the ductile fracture; there is a clear dimple-like structure, and almost all whiskers near the fracture surface break. No whiskers are pulled out, indicating the NDZ-102 coupling agent is better than the NDZ-101 coupling agent for the interfacial interaction. Furthermore, the whiskers in the composite material all have a certain orientation, indicating that in extrusion blending, the shearing action of the screw force plays a role in the orientation of the whiskers.

Advantages of dry processing are that this method is convenient and simple; and performance and microstructure analyses of the effect of the calcium carbonate whisker surface modification more intuitively and realistically reflect the interfacial action of whiskers and the matrix resin and dispersion of the whiskers in the matrix. Disadvantages are that because

(a) (b)

Figure 5.16   SEM photographs of composite material modified by coupling agents. (a) NDZ-101. (b) NDZ-102.

the amount of coupling agent is very small, dispersion tends to be uneven, and the performances of the composite material are not only related to the surface modification agent and its amount, but also to the conditions of surface treatment and the preparation of composite materials, and therefore consistent results are difficult to obtain.

5.4 Calcium Carbonate Whiskers Modified by Silane Coupling Agents

Silane coupling agents could be used for modification of various different organic functional groups on the surface of inorganic materials. The current research in this area is still very active. The common silane coupling agents for the surface modification of calcium carbonate whiskers are mainly KH-550, KH-560, and KH-570.

When treating calcium carbonate whiskers using a silane coupling agent, different researchers have different methods, but all show good surface processing performance. For example, Zhou[14] use KH-560 to modify calcium carbonate whiskers. The method is as follows. First prepare 95% (mass fraction) ethanol, 5% (mass fraction) mixed solution of deionized water. Add glacial acetic acid to adjust the pH to 4.5–5.5 and add 2 mL of the silane coupling agent KH-560 with stirring. Add 5.0 g of calcium carbonate whiskers after hydrolization for 5 minutes and continue stirring for some time. Filter, dry, and collect the sample. Infrared spectra of calcium carbonate whiskers before and after modification are shown in Figure 5.17.

Figure 5.17 shows that there is an absorption peak (α) at a wavelength of 2921 cm^{-1}, and a strong absorption peak of the C–O double bond (β) at 1500 cm^{-1}. The peak area is obtained by integrating these two absorption peaks, and the ratio of these two absorption peak area (A_α/A_β) is calculated. A_α/A_β increases probably because KH-560 is adsorbed or linked to

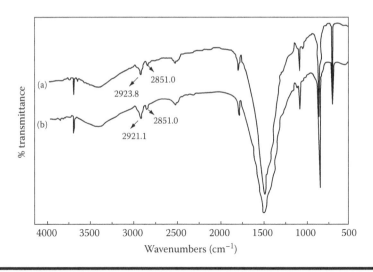

Figure 5.17   Infrared spectra of unmodified (a) and modified (b) calcium carbonate whiskers.

the surface of calcium carbonate whiskers after treatment with the coupling agent KH-560, so that absorption (α) intensity increases. On the other hand, if the coupling agent coats or link to the whisker surface, the intensity of C=O double bond absorption peaks (β) decreases, resulting in an increase of the integral area ratio of A_α/A_β after modification. Thus, silane coupling agents have certain interactions with calcium carbonate whiskers.

Juan Hao et al.[15] use KH-550 to modify calcium carbonate whiskers as follows. The silane coupling agent with a mass fraction of 0.5%–1.0% is formulated as a 95% ethanol solution, and then added to the whisker filler with stirring and dried at 120°C–160°C. Yao et al.[16] use KH-570 to treat calcium carbonate whiskers as follows. Add 1 g of KH-570 to 100 g of ethanol and then add 10 g of calcium carbonate whiskers after the silane coupling agent is dissolved. Disperse ultrasonically for 1 hour and then centrifuge for 10 minutes at 2000 rpm. Remove the supernatant, wash with ethanol, centrifuge, and remove the supernatant. Repeat three times and then dry.

5.5 Research Progress on Polymers Filled with Calcium Carbonate Whiskers

Currently, surface-modified calcium carbonate whiskers are used mainly in the following polymer materials.

5.5.1 Polypropylene

Liu et al.[17,18] used calcium carbonate whiskers to modify polypropylene (PP). The whisker-filled samples present better mechanical properties compared to the composite samples filled with ordinary light calcium carbonate. The impact strength and tensile strength increase with an increasing amount of whisker filler. When the filling content is greater than 30%, the impact strength of the material increases significantly. Compared with potassium titanate whisker-filled PP, calcium carbonate whisker-reinforced PP is slightly better. When the filling content is 30%, the tensile strength of calcium carbonate whisker-filled PP is 1.18 times that of pure PP.

Yang and He[19] found that when filled with 15% calcium carbonate whiskers, the tensile breaking strength of calcium carbonate whiskers/PP material increases by 35.7%, the flexural modulus increases by 117%, and the impact strength increases by 31.5%. Quan and Sun[20] studied the reinforcement and toughening of PP by calcium carbonate whiskers and found that the composite material filled by unmodified calcium carbonate whiskers has a better effect at a low filling content, whereas the composite material filled by modified calcium carbonate whiskers has a better combined effect at a high filling content.

5.5.2 Polyvinyl Chloride

When polyvinyl chloride (PVC) is filled by calcium carbonate whiskers, the impact strength of the composite material reaches 12.71 kJ/m², the tensile strength is up to 46.45 MPa, and the elongation at break is 4.21%, which meet the performance requirements of PVC sheets. In addition, calcium carbonate whiskers can improve anti-friction properties of the matrix material.[21] Min Wang also confirmed that the notched impact strength of the calcium carbonate whisker-filled PVC material increases by 217% compared to the composite material without calcium carbonate whiskers, and the tensile strength and flexural modulus of elasticity are also improved significantly.[22]

In addition, different coupling agents have different effects when calcium carbonate whiskers are processed. The impact strength of the composites filled by stearic acid modified calcium carbonate whiskers is better than that of the composite material filled by titanate coupling agent modified whiskers.[19]

5.5.3 Polystyrene

Sasaki et al.[23,24] developed medical capsules with a core/shell structure, which use calcium carbonate whiskers as the nucleus and crosslinked polystyrene (PS) as the shell. Urayama et al.[25] prepared inorganic filler and L-polylactide (poly-L-lactide) capsule composite material by a melt blending technique. With an increasing volume fraction of whiskers, the bending strength of the composite material declines. This is probably because calcium carbonate whiskers are broken in the blending process, and the aspect ratio decreases. Because calcium carbonate can be decomposed under acidic conditions, the advantage of using calcium carbonate whiskers to fill medical capsules is that it is decomposed by human gastric acid secretion and then eliminated. Calcium carbonate whiskers are excellent enhancing fillers with no harm to the human body. They may also be used as fillers for biodegradable plastics.

5.5.4 Polyacrylate

Inorganic whiskers of small size can also be applied as thickening agents in coating to improve viscosity, thixotropism, and surface hardness, and thus can improve paint adhesion, cementing strength, cracking resistance, scratch and scrub resistance, and other properties.[26] In addition, the good dispersion of whiskers can increase the spraying effect, make construction easier, and improve production yield. A reasonable combination with a resin system can also improve the thermostability and fire resistance of the coating.

Zhao[10] used calcium carbonate whiskers to fill epoxy acrylate and urethane acrylate coatings. When the whisker filler content was 15%, the tensile strength of both coatings reached the maximum of 13.7 MPa and 11.1 MPa respectively, improving 110% and 95% over the pure matrix resin coating. When the whisker filler content was 10%, the impact strength improved by 40% and 25%, respectively, over the pure matrix resin coatings. Wang et al.[27] compared the effects of different fillers on ethylene–vinyl acetate copolymer emulsion influence (EVA) and styrene–acrylic emulsion. The viscosity of the emulsion system filled by unmodified whiskers and light calcium carbonate is less than that of the system filled by modified calcium carbonate whiskers.

5.5.5 Polyether Ether Ketone

Asbestos is widely used in the field of friction materials and has been banned because of its toxicity. The calcium carbonate whisker is one of the ideal substitutes for asbestos because of its nontoxicity and good temperature resistance, and can significantly improve the wear resistance of materials.[6] A Japanese patent has been reported for calcium carbonate whisker-reinforced friction material, and a friction material containing calcium carbonate whiskers and glass fibers has been developed. The characteristic of this friction material is light weight, high strength, and stable friction performance.[28] Using calcium carbonate whiskers for friction material meets the requirements of high strength and security.

Lin et al.[29-32] studied the friction and wear effects of polyether ether ketone (PEEK) filled by calcium carbonate whiskers. With a gradual increase of the amount of filling whiskers, the friction coefficient of the composite continues to decrease, especially when the content of the filling is less than 15%, and the wear rate of the material drastically declines. For example, the friction coefficient is only 13.2% of pure PEEK when the whisker content is 15%. When the whisker content is 25%–30%, the material has the best friction and wear properties. Furthermore, the calcium carbonate whisker filler has a great effect on the thermal expansion coefficient of the composite. For a PEEK composite with 25% whisker filling, the average thermal expansion coefficient at 25°C–120°C is only half that of the pure polyether ether.

Li et al.[33] studied the performance of PEEK composite co-filled by calcium carbonate whiskers and polytetrafluoroethylene (PTFE) and found that under dry friction conditions, filling calcium carbonate whiskers significantly reduces the friction coefficient and wear rate of composite materials. Both properties decrease with an increasing amount of whiskers. The composite material still has very good comprehensive friction when the whisker content is 30%.

5.5.6 Nylon

Cao[34] used calcium carbonate whiskers to modify nylon-66 and obtained a composite material with a high flexural modulus of elasticity, good thermal stability, good surface finish, and dimensional stability. Compared with glass fiber filled nylon-66, a whisker-filled composite has slightly worse physical properties, but its good molding fluidity and highly smooth surface can be used in the preparation of precision parts with complex shapes (such as gears). The tensile strength of calcium carbonate whiskers/nylon-66 composite material prepared by Qingfeng Liu et al.[17] is 1.12 times that of pure PP.

5.5.7 Paper

As calcium carbonate whiskers have special directivity, high whiteness, and high filling, they can be used to fill paper. Not only can this produce inexpensive high-quality paper, but it can also make paper with special performance requirements, such as flame-retardant paper used for interior decoration, and so forth. It shows excellent printability compared with granular calcium carbonate filled paper, and reveals great application prospects.

Calcium carbonate whiskers and conventional calcite filler are added to the paper in the same way. The paper filled with whiskers has a higher retention rate and better tensile strength, tear and burst strength, and so forth.[35]

5.5.8 Other Materials

Calcium carbonate whiskers can significantly improve the anti-friction performance of a polyoxymethylene matrix composite. When the whisker content is 12%, the friction and wear rate of the composite are minimum under either water lubrication conditions or dry conditions.[36]

Wu et al.[37] compared the friction and wear properties of phenolic resin-filled calcium carbonate whiskers and potassium titanate whiskers. They found that calcium carbonate whisker-filled friction material shows the lowest sensitivity to speed and easy to cause surface abrasion whereas potassium titanate whiskers-filled friction material shows the lowest sensitivity to load and the highest sensitivity to speed.

References

1. Wu Li. *Inorganic whisker.* Beijing: Chemical Industry Press, 2005.
2. Kewei Hu, Dongsheng Li, Hui Zhong. Preparation and prospect of calcium carbonate whisker as filler. *Guangzhou Chemistry,* 31(3):57–63, 2006.

3. Jiuxing Xiang, Qiuju Sun, Shiwei Wu et al. Advances in the application of calcium carbonate whiskers. *Fine and Specialty Chemicals*, 18(1):27–30, 2010.
4. Wenyu Shang, Darong Xie, Qingfeng Liu et al. Properties of polymer material filled with aragonite whiskers. *China Plastics*, 14(3):24–27, 2000.
5. Zhaoyu Xu. Research progress of whisker and its application. *Technology & Development of Chemical Industry*, 34(2):11–17, 2005.
6. Chenying Ma. Whisker material and its application to plastics. *Modern Plastics Processing and Applications*, 9(2):60–64, 1996.
7. Litao Li, Ming Zhu, Bolong Yao et al. Study and application of $CaCO_3$ whisker on performance of drum brake linings. *New Chemical Materials*, 36(10):97–100, 2008.
8. Kewei Hu, Dongsheng Li, Hui Zhong. Preparation and prospect of calcium carbonate whisker as filler. *Guangzhou Chemistry*, 31(3):57–62, 2006.
9. Chengyin Ma. Calcium carbonate whisker reinforced material plastic. *New Chemical Materials*, 21(9):28–31, 1995.
10. Ziqian Zhao. *Preparation and application of functional UV-curing coating with $CaCO_3$ whisker.* Wuxi: Jiangnan University, 2007.
11. Xiaoming Cui. Research and application progress of inorganic whisker. *Fine Chemical Industrial Raw Materials & Intermediates*, 9(5):25–30, 2007.
12. Lixia Li, Yuexin Han, Shijie Tao. Research on surface modification of calcium carbonate whiskers. *Industrial Minerals & Processing*, 44(5):4–8, 2008.
13. Dezhen Wu. Research on NDZ-101 titanium coupling agent of polypropylene—Calcium carbonate system. *Jiangsu Chemical Industry*, 9(3):19–26, 1981.
14. Cui Zhou. *Study on application of soybean protein isolate in material areas.* Wuxi: Jiangnan University, 2009.
15. Juan Hao, Hao Liu, Lingqi Li et al. Study on epoxy resin adhesive modified by $CaCO_3$ whiskers. *China Adhesives*, 20(8):29–32, 2011.
16. Bolong Yao, Tonghua Yang, Kan Luo et al. Study and application of UV-curing composite modified by $CaCO_3$ whisker. *Engineering Plastics Application*, 35(12):29–32, 2007.
17. Qingfeng Liu, Desheng Wang, Wenyu Shang et al. Preparation of $CaCO_3$ whisker and characteristics of PP reinforced by whisker. *China Plastics Industry*, 28(1):5–9, 2000.

18. Qingfeng Liu, Zhuo Wang, Wenyu Shang. Preparation of calcium carbonate whisker. *China Journal Inorganic Chemicals Industry*, 32(2):11–12, 2000.
19. Mingshan Yang, Chenghan He. Preparation of calcium carbonate whiskers and modification of PP with it. *Modern Plastics Processing and Applications*, 20(6):21–24, 2008.
20. Xuejun Quan, Zhifu Sun. Preparation and filling performance of calcium carbonate whiskers. *Journal of Wuhan University of Technology*, 25(8):8–12, 2003.
21. Litao Su. *Study and application of reinforced and wear-resistant Whiskers*. Wuxi: Jiangnan University, 2008.
22. Min Wang, Haikui Zhou, Jianfeng Chen et al. Preparation of CaCO$_3$ whiskers by high-gravity method and their application in PVC. *Metal Mine*, 40(1):48–51, 2005.
23. Takashi Sasaki, Kttagawa Takayki, Sato Shintaro et al. Core/shell and hollow polymeric capsules prepared from calcium carbonate whisker. *Polymer Journal*, 37(6):434–438, 2005.
24. Takashi Sasaki, Shoko Kawagoe, Hajime Mitsuya et al. Glass transition of crosslinked polystyrene shells formed on the surface of calcium carbonate whisker. *Journal of Polymer Science B: Polymer Physics*, 44(17):2475–2485, 2006.
25. Hiroshi Urayama, Chenghuan Ma, Yoshiharu Kimura. Mechanical and thermal properties of poly(l-lactide) incorporating various inorganic fillers with particle and whisker shapes. *Macromolecular Materials and Engineering*, 288(7):562–568, 2003.
26. Lixia Li, Yuexin Han, Wanzhong Yin et al. The application and preparation technology of calcium carbonate whisker. *Non-Ferrous Mining and Metallurgy*, 21(S):73–74, 2005.
27. Huili Wang, Juanjuan Yang, Bin Liu et al. Application of calcium carbonate whisker in the coating. *Paint & Coatings Industry*, 34(4):52–54, 2004.
28. Masato Komatsu. Asbestos-free composite friction materials. JP05105867.
29. Youxi Lin, Chenghui Gao, Zhifang Li. Effect of CaCO$_3$ whisker content on friction and wear of filled PEEK composites. *Tribology*, 26(5):448–451, 2006.
30. Youxi Lin, Chenghui Gao. Effect of CaCO$_3$ whisker treated with SPEEK on friction and wear of filled PEEK composites. *Lubrication Engineering*, 32(11):65–68, 2007.

31. Youxi Lin, Chenghui Gao, Ning Li. Influence of CaCO$_3$ whisker content on mechanical and tribological properties of poly(ether ether ketone) composites. *Journal of Material Science Technology*, 22(5):584–588, 2006.
32. Youxi Lin, Chenghui Gao, Minghui Chen. Thermomechanical properties and tribological behavior of CaCO$_3$ whisker reinforced polyether ether ketone composites. *Advanced Tribology*, pp. 319–312. Berlin: Springer, 2010.
33. Zhifang Li. *Tribological behaviours of CaCO$_3$ whisker filled PEEK composites*. Fuzhou: Fuzhou University, 2005.
34. Youming Cao. The preparations and applications of calcium carbonate whisker. *New Chemical Materials*, 26(10):23–25, 1998.
35. Li Zhang, Xingyong Liu, Fengying Mao. Study on the property of paper filled by aragonite CaCO$_3$ whisker. *Journal of Sichuan University of Science & Engineering (Natural Science Edition)*, 20(1):73–75, 2007.
36. Yuncheng Feng. *Friction and Wear Properties of CaCO$_3$ Whisker-Filled POM Composites*. Chengdu: Xihua University, 2009.
37. Xunkun Wu, Changsong Wang, Xin Feng. Effect of whiskers-like fillers on the phenolic resin-based friction material. *Lubrication Engineering*, 32(11):122–126, 2007.

Chapter 6

Research on Polypropylene Filled with Calcium Carbonate Whiskers

Polypropylene (PP), one of the five general plastics, is among the synthetic resins with the fastest development, highest production, most brand names, and widest applications, and is widely used in all kinds of fields, such as automobiles, everyday items, furniture, and packing. The advantages of PP include abundant raw material sources, low price, easy processing, low density, excellent mechanical performance, good electrical insulation, low dielectric constant, stress cracking resistance, chemical resistance, and so on. However, it also has shortcomings, such as poor cold resistance, low-temperature brittleness, large dimension shrinkage of products, and poor compatibility with other polar polymers and inorganic fillers, which limits its application scope and functional development. Modifications through a variety of physical and chemical methods therefore are needed to prepare high-performance and functional PP materials, improve usage performance, broaden application fields, and create more economic value, and this has become a hot topic in both academia and businesses.[1]

Modification methods of PP include chemical and physical modification.

Chemical modifications mainly refer to the introduction of other components into the PP macromolecular chain through grafting, blocking, and copolymerization methods, or the change in the types of atoms or radicals in the PP molecular chain and their combination (chemical composition, molecules formed, regularity of the structure, length and distribution of the molecular chain) through crosslinking with crosslinking agents or modification using a nucleating agent and a foaming agent, to give the material high shock resistance, excellent heat resistance, and aging resistance. Commonly used chemical modification methods include copolymerization, crosslinking, chlorination, grafting, and so forth.[2] For example, propylene copolymerizes with a vinyl monomer after homopolymerization. When 2%–3% vinyl monomers are blocked, the prepared ethylene propylene copolymer rubber possesses the advantages of both polyethylene and PP; it can resist −30°C low-temperature impact; and it is widely used in automobile and industrial components and home appliances.[3] Maleic anhydride (MAH) molecules contain an anhydride groups with a double bond and other groups. The double bond is easy to open under the action of an initiator and grafts with PP molecular chains; the anhydride group can be involved in chemical reactions as a functional group. Therefore, the polarity and reactivity of PP can be improved by grafting MAH into the PP molecular chain through a melting reaction with dicumyl peroxide used as an initiator.[4]

In physical modification, the molecular aggregation structure of PP is changed by the addition of another inorganic material, an organic material, or other additives with special functions into the PP matrix, with thorough mixing. The methods rely on physical properties, deformation characteristics, and the morphological change of different components to improve the material

properties or obtain some new functions and to prepare composite materials with outstanding performance.[5] Physical modification can be achieved in general enterprises because of its
simple equipment requirements and easy operation. Physical
modifications mainly include the following methods.

1. *Filling modification.* Some organic or inorganic fillers are
   filled into PP to improve the performance of PP or reduce
   cost. Various inorganic fillers are used in PP modification
   research, including calcium carbonate, talcum powder,
   mica powder, wollastonite, and so forth. According to the
   literature reports,[6,7] to reduce costs, plastic filling enterprises in China mainly use calcium carbonate as a mineral filler, and masterbatch products are used in 90% of
   masterbatch products are used in filling polyolefin resin,
   and high-performance and high value-added products
   account for about 10%.
2. *Reinforcing modification.* Composite materials are prepared by adding some reinforcing materials into PP, such
   as glass fiber, asbestos fiber, carbon fiber, and natural
   plant fiber, to replace engineering plastics or change the
   morphological structure of polymers through special
   processing and molding methods and special molds. For
   example, Cui and Cao[8] prepared long fiber-reinforced
   PP grains using a self-designed dipping mold. The bending strength of reinforced PP increases with an increase
   of glass fiber content. At 80°C, the bending strength of
   long fiber-reinforced PP with 20%–30% glass fiber content increases 40% over that of short fibers with the same
   quality content. The strength increases 65% when the
   fiber content is above 40%.
3. *Blending modification.* Blending modification can
   improve the thermal stability, shock resistance (especially low-temperature impact resistance), and toughness
   of PP and extend the application fields of PP. Blending
   modification of PP is currently the most widely used

modification technology, including PP–rubber blending and blending of PP and other plastic resins.[9] For example, transparent PP/SEBS blends can be obtained by melt blending of propylene and hydrogenated polystyrene–butadiene–block copolymer of polystyrene (SEBS). The crystallinity of PP decreases with an increase of SEBS content; the morphology of PP/SEBS blend changes from an island structure to a double continuous phase structure, and the transparency and low-temperature impact strength of the blend increase greatly.[10]

4. *Nucleating agent modification.* Nucleating agent modification refers to the addition of the proper nucleating agent to PP to change its crystalline form, crystallization behavior, crystallization rate, and crystal structure so as to achieve material modification.

PP is a semicrystalline polymer whose physical, optical, and mechanical properties are determined to a great degree by its crystallinity, crystal form, and crystal structure. Under different crystallization conditions, it can form α, β, γ, δ, and quasi-hexagonal crystalline forms, among which α and β are the most common.

The α crystal form of PP belongs to the monoclinic system, the most common and stable crystal form, and is formed mainly under general crystallization conditions, with a melting point of 167°C and density of 0.936 g/cm³. The β crystal form belongs to the trigonal system, the melting point is 150°C, and the density is 0.922 g/cm³. Because the internal arrangement of the β crystal form is much more scattered than that of the α crystal form, the β crystal form has better impact absorption, and the impact strength of β PP is higher than that of α PP by one to two times. However, the β crystal form is thermodynamically metastable and is hardly generated in dynamics (kinetically), and the α crystal form is usually formed under melting crystallization conditions. But the β crystal grows faster than the α crystal in the main crystallization temperature range of β crystals; thus PP with a high β crystal content can be obtained by controlling crystalli-zation conditions, adding β crystal nucleation agent, shear induc-tion, oscillation, and so on. Among these methods, adding a β crystal nucleating agent is the most studied, and also is currently the only way to improve β crystal content in industry.[11]

Among numerous inorganic whiskers, calcium carbonate whiskers not only have the most advantages, but also have a high degree of whiteness, rich source of raw material, and low price, and thus are expected to become a new additive in plastic modification. According to the literature,[12] filling calcium carbonate whiskers into polymer materials can improve not only the mechanical properties and processing performance of the system, but also the hardness, friction performance, and thermal stability of the composites. The research on calcium carbonate whisker-filled PP is discussed in terms of the inter-nal mixing-compression molding method, extrusion-molding method, and extrusion-injection molding method.

6.1 Preparation of Calcium Carbonate Whisker-Filled PP Composites

6.1.1 Internal Mixing-Compression Molding

Dried PP is added into the mixing chamber of an internal mixer and is melted for 7 minutes at a speed of 40 rpm at 210°C. Then treated calcium carbonate whiskers are added and mixed for 8 minutes and the dough material is removed. The dough mate-rial is put into a clamping plate and heated until melted, then is hot molded in a plate vulcanizing machine for 5 minutes at 200°C and 11.5 MPa, and finally demolded by cooling.

6.1.2 Extrusion Molding

Dried PP and treated calcium carbonate whiskers in different proportions are mixed in a high-speed mixing machine, and then granulated through a twin-screw extruder. The granules are molded in a plate vulcanizing machine for 5 minutes at 200°C and 11.5 MPa, then demolded by cooling.

6.1.3 Extrusion-Injection Molding

Dried PP and treated calcium carbonate whiskers in different proportions are mixed in a high-speed mixing machine, granulated through a twin-screw extruder, and then injection molded.

6.2 Performance Test Methods of Composite Materials

6.2.1 Thermal Properties

The thermal properties of materials are measured using differential scanning calorimetry (DSC) with a heating rate and cooling rate of 5, 10, 15, and 20°C/minute, respectively; test temperature range of 30–220°C; N_2 as protective gas; and disposable aluminum crucible. Thermal stability is analyzed using thermogravimetric analysis with a heating rate of 20°C/minute, test temperature range of 40–550°C, N_2 as the protective gas, and disposable aluminum crucible.

6.2.2 Crystallization Properties

X-ray diffraction uses copper target radiation, with a scan range of 5–50° and a scan rate of 1.2°/minute.

6.2.3 Mechanical Properties

Tensile properties are measured in accordance with the GB/T 1040.1-2006 "Determination of plastic tensile properties" standard test. The sample is in dumbbell shape, the sample size is

115 mm × 6.5 mm × 2 mm, and the stretching rate is 50 mm/minute. Bending performance is measured according to the GB/T 9341-2000 "plastic bending performance test method" standard test. The sample size is 40 mm × 25 mm × 2 mm and the bending rate is 1 mm/minute.

6.2.4 Microstructure

The fractures of samples in the tensile test and bending test are observed using scanning electron microscopy (SEM) after metal spraying.

6.3 Performance Analysis of Composite Materials Prepared Using Extrusion Molding

6.3.1 Structure Analysis of Composite Materials

The prepared calcium carbonate whisker-filled composites are analyzed by x-ray diffraction. The results are shown in Figure 6.1.

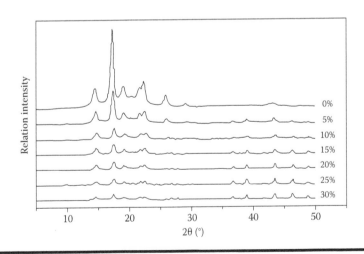

Figure 6.1   X-ray diffraction chart of composite materials.

We can see from Figure 6.1 that the PP materials before and after filling with calcium carbonate whiskers all show five diffraction peaks at 2θ diffraction angles of 14.4°, 17.2°, 18.9°, 21.6°, and 22.2°, which correspond to the diffractions of the α crystal on (110), (040), (130), and overlapping (131), (111) crystal planes.[11] In addition, the five diffraction peaks at 2θ diffraction angles of 36.6°, 38.9°, 43.4°, 46.3°, and 48.8° correspond to the diffraction peaks of calcium carbonate whiskers. With an increase of whisker content, the intensity of diffraction peaks of the α crystal gradually decrease, while the intensity of diffraction peaks of calcium carbonate whiskers gradually increase, indicating that the filling of calcium carbonate whiskers does not change the crystal form of PP, but influences the crystallization performance of PP.

Crystallinity can be calculated using the corresponding x-ray diffraction peak area. The calculation formula of crystallinity is as follows:

$$X_c = (A_c + KA_\alpha) \times 100\% \tag{6.1}$$

where
    A_c = the diffraction peak area of the crystal phase in the diffraction intensity curve
    A_α = the diffraction peak area of the amorphous phase in the diffraction intensity curve
    K = the correction factor

6.3.2 Thermal Performance Analysis of Composite Materials

6.3.2.1 DSC Analysis

Thermal analysis is a commonly used analytical method in the study of thermal properties of polymer materials. The thermodynamic properties of PP materials before and after filling with whiskers are analyzed using DSC with both a heating rate and a cooling rate of 20°C/minute. The DSC heating curves of PP composites filled with whiskers with different filling content are shown in Figure 6.2; the cooling curves are shown in Figure 6.3; and the thermal property data are listed in Table 6.1.

In Table 6.1, T_m stands for the maximum temperature of the melting peak in the heating curve; ΔH_m stands for melting enthalpy; T_c stands for the maximum temperature of the

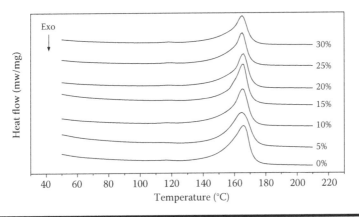

Figure 6.2 DSC heating curves of composite materials with different whisker filling content.

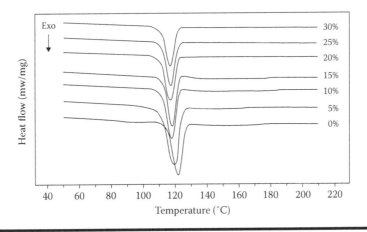

Figure 6.3 DSC cooling curves of composite materials with different whisker filling content.

Table 6.1 Thermal Property Data of Composite Materials with Different Filling Contents

Filling Content (%)	T_m (°C)	$\Delta H_m/$ (J/g)	T_c (°C)	$\Delta H_c/$ (J/g)	T_{co} (°C)	T_{cq} (°C)	ΔT (°C)
0	166.0	81.3	121.6	82.4	129.0	102.1	26.9
5	164.7	72.5	118.8	84.8	127.3	97.4	29.9
10	165.4	69.3	117.7	75.2	124.9	100.3	24.6
15	165.6	66.8	118.2	70.8	124.7	100.2	24.5
20	165.7	60.3	117.1	66.2	124.7	101.2	23.5
25	164.9	57.9	117.2	60.9	124.5	100.5	24.0
30	165.0	54.7	116.8	58.2	124.0	100.5	23.5

crystallization peak in the cooling curve; and ΔH_c stands for crystallization enthalpy. As can be seen from Table 6.1, when PP is filled with calcium carbonate whiskers, the T_m of the composite materials changes slightly; T_c decreases as a whole; ΔH_m gradually decreases with increasing whisker content; and ΔH_c first increases and then decreases with the addition of whiskers and reaches the maximum value of 84.8 J/g when the filling content is 5%.

The width of the crystallization peak is represented by $\Delta T = T_{co} - T_{cq}$, where T_{co} stands for the initial temperature of the cooling curve deviating from the baseline when crystallization begins; T_{cq} stands for the temperature when the cooling curve overlaps with the baseline, that is, when crystallization ends. The results are listed in Table 6.1. T_{co} of pure PP is 129°C and decreases with an increase of whisker filling content. T_{co} drops to 127.3°C when PP is filled with 5% whiskers; T_{co} drops to 124.9°C when the filling content is 10%; after that, T_{co} changes a little with an increase of whisker filling content, suggesting that the initial crystallization temperature of PP composite materials decreases after filling with calcium carbonate whiskers. It is generally believed that the crystallization of PP requires the rotation and folding of chain segments. This is a relaxation process, and discharging into the crystal lattice requires some time. Crystallization under low temperature is conducive for macromolecular chains or chain segments entering the crystal lattice; however, the poor activity leads to great differences in crystallization maturity.

As can be seen from the ΔT calculation results in Table 6.1, ΔT of pure PP is 26.9°C; ΔT increases to 29.9°C after filling with 5% whiskers, and then quickly declines to 24.6°C when the filling content is 10%. After that, ΔT changes little as the amount of filled whiskers continues to increase, indicating that the crystallization peak of PP widens when filled with a small amount of calcium carbonate whiskers, and the crystallization peak of PP narrows with increasing whisker content.

Like the crystallization process of small molecules, the crystallization of polymers includes the formation of a crystal nucleus and the growth of the crystal. The nucleation method includes homogeneous nucleation and heterogeneous nucleation. In homogeneous nucleation, the nucleus forms when polymer chains or segments with low thermal motion energy in melts form orderly arrangements of chain beams with thermodynamic stability. In heterogeneous nucleation, the nucleus forms when foreign impurities, incompletely fused residual polymer crystals, dispersed solid particles, or the container wall adsorb polymer chains or segments in melts and form orderly arrangements. Once the crystal nucleus forms, the polymer chain will spread and form orderly arrangements through the nucleus and cause the crystal grain to grow.

The structure regularity of the polymer chain is just the internal reason for the crystallization ability of polymer. A polymer with crystallization ability can crystallize only under proper conditions, and this is the external cause of polymer crystallization. The crystallization rate of a polymer depends on temperature, pressure, impurity, solvents, the structure of the polymer chain itself, and a variety of other factors. The dependence of the crystallization rate on temperature is determined by the dependence of the nucleation rate and crystal growth rate on temperature. The dependence of the nucleation rate on temperature is related to the nucleation mechanism. Heterogeneous nucleation can occur at a high temperature, whereas homogeneous nucleation occurs only at a low temperature.[12]

To further study the influences of calcium carbonate whiskers on the crystallization performance of PP, we studied the crystallization dynamics of pure PP and 5% whisker-filled PP. The dynamic DSC curves of these two materials are determined, as shown in Figures 6.4 and 6.5, respectively. The numbers in the figures represent cooling rates.

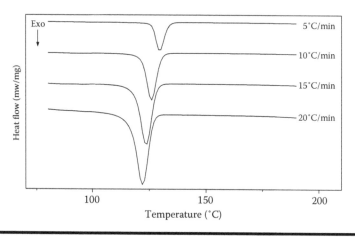

Figure 6.4   Dynamic DSC cooling curves of pure PP.

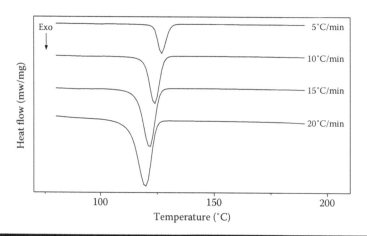

Figure 6.5   Dynamic DSC cooling curves of 5% whisker-filled PP.

The nonisothermal crystallization kinetics of polymers can start with isothermal crystallization, and be corrected considering the characteristics of nonisothermal crystallization. The common DSC methods include Jeziorny, Ozawa, and MoZhishen methods.[11] The nonisothermal crystallization kinetics of pure PP and 5% whisker-filled PP are compared using the Jeziorny modifying method of the Avrami equation.

Usually, the Avrami equation is applied in isothermal crystallization kinetics[13]:

$$[1 - X(t)] = \exp[-Zt^n] \tag{6.2}$$

where
  $X(t)$ = relative crystallinity at time t
  Z = isothermal crystallization rate constant
  n = Avrami index

The value of n can reflect the growth mechanism of polymers and crystallization nucleation, with $n = 1$ or 2 corresponding to one-dimensional fiber growth; $n = 2$ or 3 corresponding to two-dimensional sheet growth; $n \geq 3$ corresponding to three-dimensional spherocrystal growth. That is to say, the larger the n value, the higher is the dimensional value of crystal grains during growth.

In the process of nonisothermal crystallization, the relative crystallinity $X(T)$ can be calculated according to the following equation:

$$X(T) = \frac{\int_{T_{co}}^{T} (dH/dT)dT}{\int_{T_{co}}^{T_{cq}} (dH/dT)dT} \tag{6.3}$$

where
  T_{co} = the onset temperature of crystallization
  T_{cq} = the end temperature of crystallization
  dH/dT = the heat flow rate

Figures 6.6 and 6.7 show the relationships of relative crystallinity $X(T)$ and temperature T of pure PP and 5% filled PP at different cooling rates. The reverse S curves in Figures 6.6 and 6.7 show the nucleation stage with a slower crystallization rate, the faster initial crystallization stage, as well as the relatively slow secondary crystallization stage.[11]

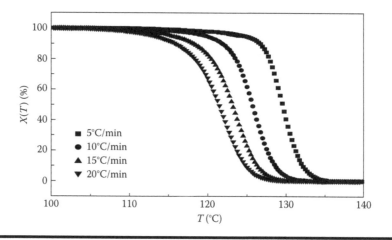

Figure 6.6 *X(T) – T* curves of pure PP at different cooling rates.

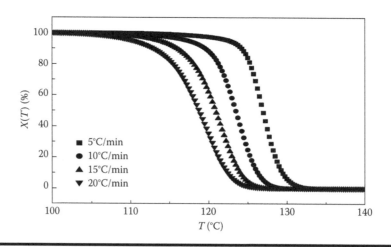

Figure 6.7 *X(T) – T* curves of 5% whisker-filled PP at different cooling rates.

The curves in Figures 6.6 and 6.7 are turned into the relationship of relative crystallinity $X(T)$ with time t.

$$t = \frac{T_{co} - T}{\Phi} \qquad (6.4)$$

where
    T = the temperature at time t
    T_{co} = the crystallization beginning temperature during the
        cooling process
    Φ = the cooling rate

Figures 6.8 and 6.9 are the $X(T) - T$ relationship diagrams of pure PP and 5% whisker-filled PP at different cooling rates. It can be seen that the crystallization rates of both pure PP and 5% whisker-filled PP increase with an increasing cooling rate. That is to say, the higher the cooling rate, the higher the crystallization rate of PP.

The Avrami equation should be modified when dealing with nonisothermal crystallization kinetics.

Jeziorny defined[14]

$$\lg Z_C = \lg Z/\Phi \qquad (6.5)$$

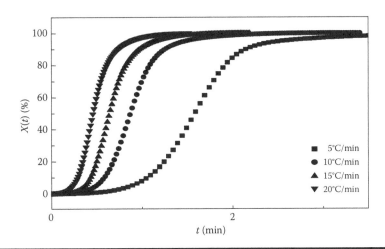

Figure 6.8 $X(t) - t$ curves of pure PP at different cooling rates.

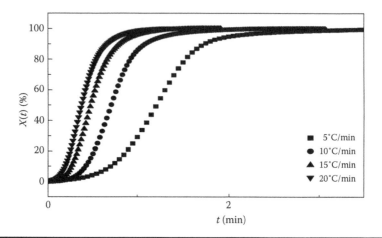

Figure 6.9 $X(t) - t$ **curves of 5% whisker-filled PP at different cooling rates.**

and applied the Avrami equation to the DSC curve with constant speed and varying temperature.

Based on the corrected results, the $\lg\{-\ln[1 - X(t)]\}$ and $\lg t$ linear relationship diagrams can be obtained after the non-linear parts at the beginning and the end of crystallization are removed, as shown in Figures 6.10 and 6.11.

As shown in Figures 6.10 and 6.11, linear fitting is rational, suggesting that it is feasible to use the Jeziorny method to treat

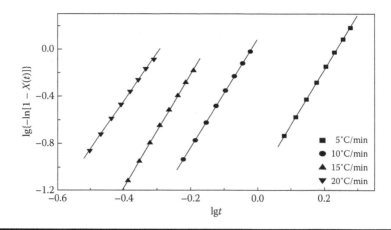

Figure 6.10 $\lg\{-\ln[1 - X(t)]\} - \lg t$ **curves of pure PP at the different cooling rates.**

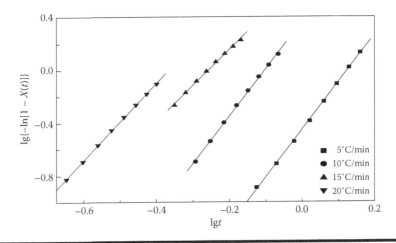

Figure 6.11 lg{−ln[1 − X(t)]} − lg*t* curves of 5% filled PP at different cooling rates.

the nonisothermal crystallization process of filled PP at a given cooling rate. The fitting line slope is *n*, the intercept is lg*Z*, and the nonisothermal crystallization rate constants Z_C can be calculated using formula (6.5), as listed in Table 6.2.

$t_{1/2}$ is the half-crystallization time, that is the time of 50% crystallization. $t_{1/2}$ can be used to represent the crystallization rate; the smaller the $t_{1/2}$, the faster is the crystallization rate.

$$t_{1/2} = (\ln 2/Z)^{1/n} \tag{6.6}$$

Table 6.2 Nonisothermal Crystallization Kinetic Parameters of Pure PP and 5% Filled PP

Φ/(°C/ minute)	Pure PP				5% Filled PP			
	$t_{1/2}$ (minutes)	n	Z	Z_C	$t_{1/2}$ (minutes)	n	Z	Z_C
5	1.59	4.66	0.08	0.60	1.21	3.58	0.35	0.81
10	0.88	4.63	1.24	1.02	0.71	3.59	2.32	1.09
15	0.65	4.81	5.66	1.12	0.48	2.69	4.87	1.11
20	0.47	4.08	15.53	1.15	0.38	2.88	11.06	1.13

Half-crystallization time $t_{1/2}$, Avrami index n, and crystallization rate constant Z and Z_C are listed in Table 6.2.

As can be seen from Table 6.2, at the same Φ value, $t_{1/2}$ values of 5% whisker-filled PP are all smaller than those of pure PP. Furthermore, the larger the Φ value, the smaller are the $t_{1/2}$ values of pure PP and 5% filled PP, indicating that the crystallization rates of PP are improved by the addition of calcium carbonate whiskers, and the larger the cooling rate, the faster is the crystallization rate.

In addition, the n values of pure PP are between 4.08 and 4.81, belonging to 3D growth, whereas the n values of 5% filled PP are between 2.69 and 3.59, which are smaller than the n values of pure PP, indicating that the nucleation and growth mode of whisker-filled PP have changed. Moreover, Z_C values of PP before and after filling increase with an increase of Φ values, suggesting that the faster the cooling rate, the faster is the crystallization rate. At a slow cooling rate, such as 5°C/minute, the Z_C value (0.81) of 5% whisker-filled PP and the Z_C value (0.61) of pure PP show a large difference. At a faster cooling rate, such as 20°C/minute, the Z_C value (1.15) of 5% whisker-filled PP and the Z_C value (1.13) of pure PP show a

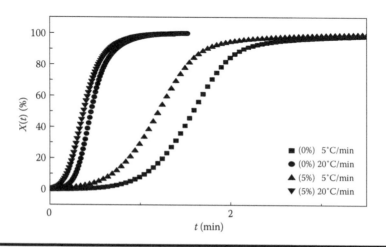

Figure 6.12 The $X(t) - t$ curves of two kinds of materials at the cooling rates of 5°C/minute and 20°C/minute.

very small difference, which is consistent with the preceding conclusion with regard to $t_{1/2}$.

Figure 6.12 shows the crystallization curves of pure PP and 5% filled PP at cooling rates of 5°C/minute and 20°C/minute. At a 5°C/minute cooling rate, the crystallization rate of whisker-filled PP is obviously faster than that of pure PP, but at a 20°C/minute cooling rate, they are similar, indicating that calcium carbonate whiskers have a greater impact on the crystallization rates of PP at slower cooling rates.

6.3.2.2 Thermal Stability Analysis

The thermogravimetric curves of pure PP and PP filled with different amounts of calcium carbonate whiskers are shown in Figure 6.13, and the thermal decomposition temperature data at 5% thermal weight loss are listed in Table 6.3.

As can be seen from Table 6.3, the 5% thermal decomposition temperatures of composites increase significantly when PP is filled with calcium carbonate whiskers. The thermal decomposition temperature is highest at 410.7°C when 25% whiskers are filled. This is because calcium carbonate whiskers have good thermal stability, which improves the thermal stability of

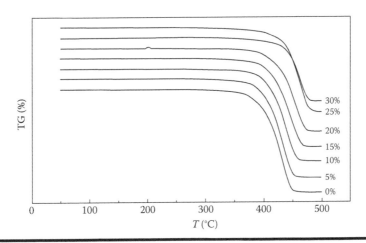

Figure 6.13 Thermogravimetric curves of pure PP and filled PP.

Table 6.3 Thermal Decomposition Temperatures of the PP before and after Filling

Filling Content (%)	0	5	10	15	20	25	30
T_d (°C)	366.9	375.8	382.5	388.8	393.5	410.7	400.6

Note: T_d is the thermal decomposition temperature at 5% thermal weight loss.

composite materials and makes use of the advantages of inorganic whiskers.

6.3.3 Mechanical Properties of Composite Materials

The stress–strain test is the most widely used mechanical test in material research. Tensile strength, breaking strength, fracture strain, elastic modulus, breaking energy, and other physical quantities can be obtained from determined stress–strain curves.

The basic characteristics of a brittle fracture are as follows: before fracture, homogeneous deformation of the sample, almost linear stress–strain curve, small fracture strain, and low fracture energy; after fracture, almost no residual strain and section perpendicular to stress direction. Generally, brittle fracture is caused by the tensile stress component of a force.

The basic characteristics of a ductile fracture are as follows: before fracture, large deformation, uneven deformation along the specimen length direction (neck shrinkage), nonlinear stress–strain relationship, zero or even negative slope of the stress–strain curve in the large deformation stage, and high fracture energy; after fracture, obvious residual strain and section not perpendicular to stress direction. Generally, a ductile fracture is caused by the shear stress component.

6.3.3.1 Tensile Properties Analysis

A tensile test is conducted according to GB/T 1040.1-2006, and tensile strength σ is calculated using the following formula:

$$\sigma = \frac{F}{A} \qquad (6.7)$$

where
  F = measured corresponding load, N
  σ = tensile strength, MPa
  A = original cross-sectional area of a sample, mm^2

Elongation at break is calculated using the following formula:

$$\varepsilon = \frac{L_1 - L_0}{L_0} \times 100\% \qquad (6.8)$$

where
  L_1 = the length of a sample when broken
  L_0 = the original length of a sample

The tensile strength and elongation at break data of the composite materials are listed in Table 6.4.

As can be seen from Table 6.4, the tensile strengths of the composite materials are slightly smaller than that of pure PP, while the elongations at break of the composite materials are all larger than that of pure PP. The elongation at break of the composite material reaches a maximum value of 14.7% when 5% calcium carbonate whiskers are filled into PP, indicating that the addition of calcium carbonate whiskers improves the anti-fracture ability of materials.

Table 6.4 Tensile Property Data of Composite Materials

Filling Content (%)	*0*	*5*	*10*	*15*	*20*	*25*	*30*
Tensile strength (MPa)	19.5	18.9	18.8	18.9	18.2	16.1	16.0
Elongation at break (%)	13.4	14.7	13.9	14.0	14.5	13.6	13.7

6.3.3.2 Bending Performance Analysis

Bending tests are conducted in accordance with the national standard GB/T 9341-2000, and bending strength is calculated using the following formula:

$$\sigma_f = \frac{3FL}{2bh^2} \qquad (6.9)$$

where

σ_f = bending strength, MPa
F = applied force, N
L = span, mm
b = specimen width, mm
h = Specimen thickness, mm

The strain force is calculated as follows:

$$\varepsilon_{fi} = \frac{6hs_i}{L^2} \qquad (6.10)$$

where

ε_{fi} = bending strain, represented by ratio or %
s_i = single deflection, that is the deviation distance of the top or bottom of the sample span center from the original position in the bending process, mm
L = span, mm
h = specimen thickness, mm

Bending strength and fracture bending strain data of composite materials are listed in Table 6.5.

As can be seen from Table 6.5, with the increase of calcium carbonate whiskers, the flexural strengths of the composites increase first and then decrease. When the whisker filling content is 5%–10%, the bending stress of the composite material increases 11.5% over that of pure PP. In addition, the bending fracture strains of composite materials filled with whiskers

Table 6.5 Tensile Property Data of Materials before and after Filling

Filling Content (%)	0	5	10	15	20	25	30
Tensile strength (MPa)	38.4	42.8	42.3	38.9	37.5	36.6	35.5
Bending fracture strain (%)	6.1	16.7	15.6	9.7	7.5	7.5	9.4

are all higher than that of pure PP. When the whisker filling content is 5%, the bending fracture strain reaches 16.7%, which is 173% higher than that of pure PP, indicating that calcium carbonate whiskers have a toughening effect on PP. This is because calcium carbonate whiskers are a single crystal fibrous material with a certain length-to-diameter (L/D) ratio. When stress is acting on composite materials, calcium carbonate whiskers dispersed in PP can locally resist strain, and therefore can buffer and reduce stress and increase bending strain.[15]

6.3.4 *Microstructure Analysis of Composite Materials*

The compatibility of calcium carbonate whiskers and matrix material and the distribution in matrix materials have a great influence on the performance of composite materials. The tensile fracture and bending fracture of composites are observed using SEM, and the SEM images of tensile fracture surface are shown in Figures 6.14 and 6.15.

As shown in Figure 6.14, the fracture of the composites filled with 5% whiskers is brittle fracture, and broken and scattered whisker fragments can be seen on the fracture surface (Figure 6.14a). With an increase of filling amount, the fracture transforms into a ductile fracture when the filling content reaches 15%, and an obvious toughening nest distribution can be seen in the SEM image, indicating that the intervention of whiskers changes the fracture property of PP (Figure 6.14b).

Figure 6.15a is the amplified tensile fracture photo of the composite material filled with 5% whiskers, and broken whiskers

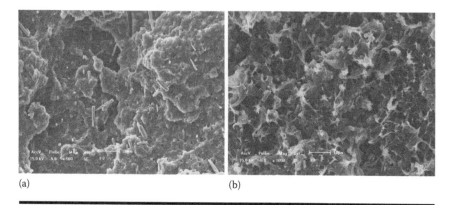

(a)　　　　　　　　　　　　　　　　(b)

Figure 6.14　SEM images of tensile fracture surfaces (×500). (a) 5% whisker content. (b) 15% whisker content.

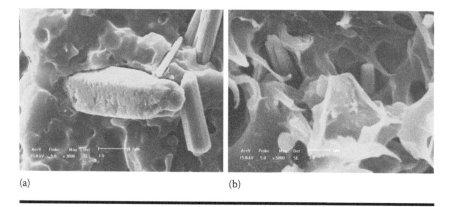

(a)　　　　　　　　　　　　　　　　(b)

Figure 6.15　SEM images of tensile fracture surfaces at different magnifications. (a) 5% whisker content (×3000). (b) 15% whisker content (×5000).

and necking deformation can be clearly seen. Figure 6.15b is the amplified tensile fracture photo of the composite material filled with 15% whiskers, and necking whiskers and cracking whiskers can be apparently seen, which indicates that whiskers must carry a force in the stretching process. This is because the strength of whiskers is much higher than the strength of the surrounding matrix. When matrix material breaks under a stretching effect, it generates a strong force on the interface of the matrix material and whiskers, and under this effect, whiskers are pulled off the matrix or break up, thus absorbing energy and improving the strength of the material. But because whiskers are short, their pulling effect is much smaller than that of long fibers.

SEM images of the bending fracture surface are shown in Figures 6.16 and 6.17. As can be seen from Figures 6.16 and 6.17, the bending fractures are all ductile fractures. Scattered and broken calcium carbonate whiskers can be seen on the fracture surface, and large and deep toughening nests appear on the bending fracture, which indicates that the material produces a large plastic deformation in the breaking down process. When the front ends of macro cracks are inside a sample or micro cracks develop in the micro zone containing

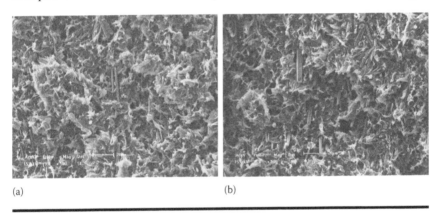

(a) (b)

Figure 6.16 **SEM images of bending fracture surfaces (×500). (a) 5% whisker content. (b) 15% whisker content.**

(a) (b)

Figure 6.17 **SEM images of bending fracture surfaces (×2000). (a) 5% whisker content. (b) 15% whisker content.**

whiskers, they can continue to expand only after whiskers are pulled out or broken, so whiskers in the matrix have the effect of preventing crack propagation and accelerating energy dissipation, and therefore improve the toughness of composite materials. However, whiskers are very short, so their pulling effect is much less obvious than the effect of long fibers.[16,17]

6.4 Performance Analysis of Composite Materials Prepared Using Extrusion Molding

6.4.1 Structure Analysis of Composite Materials

X-ray diffraction tests are conducted on composite materials and the results are shown in Figure 6.18.

As we can see from Figure 6.18, PPs before and after filling all have α crystal forms, with 2θ diffraction angles at 14.8°, 17.6°, 19.3°, 22.0°, and 22.6° respectively, which correspond to diffractions of (110), (040), (130), and overlapping (131), (111) crystal planes. With an increase of filled calcium carbonate whiskers, the crystal structure of PPs does not change, but the intensity of five diffraction peaks decreases. Furthermore, the diffraction peaks of calcium carbonate whiskers are located

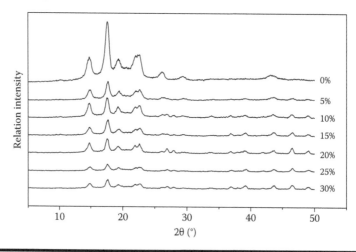

Figure 6.18   X-ray diffraction diagram of composite materials.

at 36.9°, 39.2°, 43.7°, 46.6°, and 49.1°, and the intensity of the diffraction peaks increases with increasing whisker content, indicating that calcium carbonate whiskers in composite materials are affected by the crystallization properties of PP.

6.4.2 Thermal Performance Analysis of Composite Materials

6.4.2.1 DSC Analysis

The DSC heating curves of pure PP and composite materials filled with different amounts of calcium carbonate whiskers are shown in Figure 6.19. The cooling curves are shown in Figure 6.20; both the heating rate and cooling rate are 20°C/minute. The thermal performance data are shown in Table 6.6.

As can be seen from Table 6.6, before and after filling with calcium carbonate whiskers, the melting temperature T_m of composite materials shows little change, but the melting enthalpy T_c decreases significantly; and all melting enthalpy ΔH_m and crystallization enthalpy ΔH_c decrease with increasing whisker filling content. In addition, T_{co} of pure PP is 130.4°C, T_{co} of PP filled with 5% whiskers decreases to 125.2°C, and

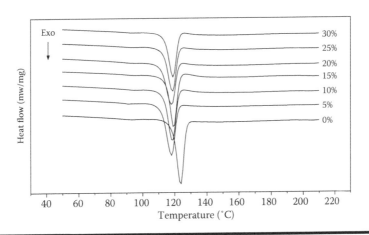

Figure 6.19 DSC heating curves of PP materials filled with different amounts of calcium carbonate whiskers.

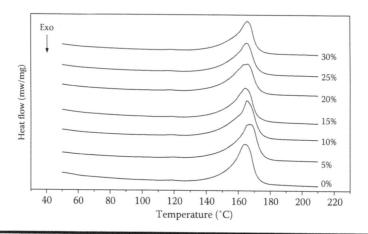

Figure 6.20 DSC cooling curves of PP materials filled with different amounts of calcium carbonate whiskers.

after that T_{co} shows little change, indicating that the crystallization temperature of calcium carbonate whisker-filled PP decreases. The calculated ΔT data show that the crystallization peaks of PPs become wider when filled with larger amounts of whiskers.

The crystallization kinetics of pure PP and 5% filled PP were studied. The dynamic DSC curves of two materials were measured, and the cooling curves are shown in Figures 6.21

Table 6.6 Thermal Analysis Data of PPs before and after Filling

Filling Content (%)	T_m (°C)	ΔH_m (J/g)	T_c (°C)	ΔH_c (J/g)	T_{co} (°C)	T_{cq} (°C)	ΔT (°C)
0	164.2	77.43	123.7	84.88	130.4	107.7	22.7
5	167.7	76.67	117.9	83.84	125.2	102.8	22.4
10	165.4	76.62	118.5	79.44	125.5	103.0	22.5
15	164.1	65.50	119.3	77.65	125.9	103.1	22.8
20	165.8	67.18	118.1	73.02	124.5	100.4	24.1
25	165.4	64.65	118.7	69.58	125.1	102.0	23.1
30	165.7	61.38	119.1	66.99	125.3	101.8	23.5

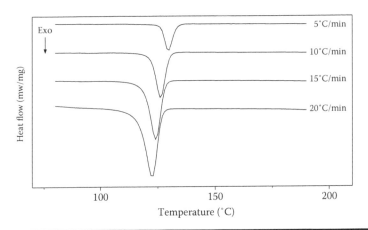

Figure 6.21   Dynamic DSC cooling curves of pure PP.

and 6.22 respectively, where the numbers stand for cooling rates.

The nonisothermal crystallization kinetics of pure PP and 5% filled PP were studied using the Avrami equation corrected by the Jeziorny method.

Avrami equation:

$$[1 - X(t)] = \exp[-Zt^n] \quad \text{same as (6.2)}$$

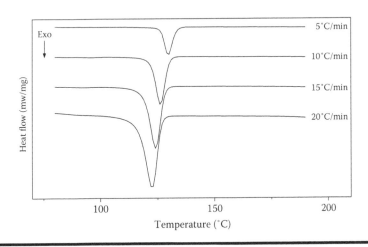

Figure 6.22   Dynamic DSC cooling curves of 5% whisker-filled PP.

where

$X(t)$ = The relative crystallinity at time t

Z = isothermal crystallization rate constant

n = Avrami index

Figures 6.23 and 6.24 show the relationship diagrams of relative crystallinity $X(T)$ with temperature T of pure PP and 5% filled PP at different cooling rates. The calculated method is the same as in Equation 6.3.

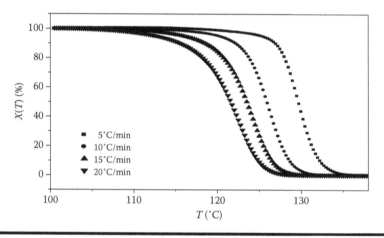

Figure 6.23 $X(T) - T$ curves of pure PP at different cooling rates.

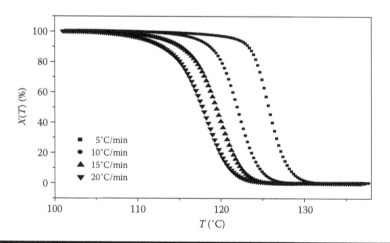

Figure 6.24 $X(T) - T$ curves of 5% whisker-filled PP at different cooling rates.

The reverse S curves in Figures 6.23 and 6.24 show the nucleation stage with a relatively slow crystallization rate, the relatively faster initial crystallization stage, and relatively slow secondary crystallization stage.

Figures 6.25 and 6.26 are the $X(t) - t$ relationship diagrams of pure PP and 5% filled PP respectively at different cooling rates.

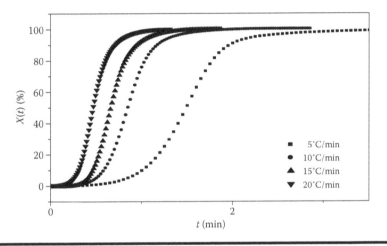

Figure 6.25 $X(t) - t$ **curves of pure PP at different cooling rates.**

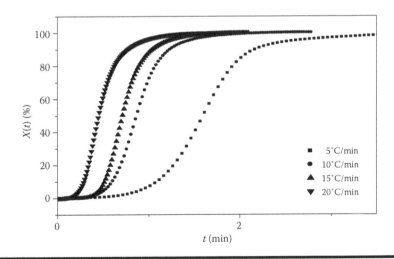

Figure 6.26 $X(t) - t$ **curves of 5% whisker-filled PP at different cooling rates.**

It can be seen from Figures 6.25 and 6.26 that the crystallization rates of both pure PP and 5% whisker-filled PP increase with an increase of cooling rates. That is to say, the faster the cooling rates, the faster are the crystallization rates of PP.

The Avrami equation is directly applied in constant heating and cooling DSC curves. The linear relationships between $\lg\{-\ln[1 - X(t)]\}$ and $\lg t$ are obtained by removing the nonlinear parts at the beginning and the end of crystallization, as shown in Figures 6.27 and 6.28.

As shown in Figures 6.27 and 6.28, linear fitting is quite ideal, suggesting that it is feasible to use the Jeziorny method to treat the nonisothermal crystallization process of PP at given cooling rates. The fitting line slope in Figures 6.27 and 6.28 is n, the intercept is $\lg Z$, and the nonisothermal crystallization rate constant Z_C is calculated according to $\lg Z_C = \lg Z/\Phi$. The half-crystallization time $t_{1/2}$, Avrami index n, rate constant Z, and Z_C are listed in Table 6.7.

We can see from the data in Table 6.7 that the greater the cooling rate Φ, the smaller is the $t_{1/2}$ of pure PP and 5% filled PP, suggesting a faster crystallization rate. At each given Φ

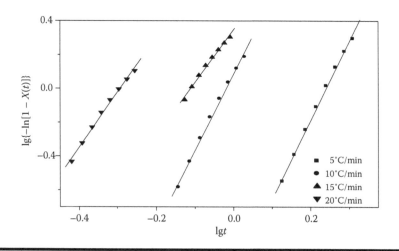

Figure 6.27 $\lg\{-\ln[1 - X(t)]\}$ – $\lg t$ curves of pure PP at different cooling rates.

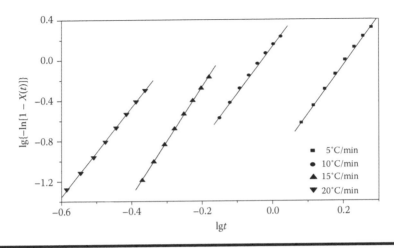

Figure 6.28 **lg{−ln[1 − X(t)]} − lgt curves of 5% whisker-filled PP at different cooling rates.**

value, the $t_{1/2}$ value of whisker-filled PP is slightly smaller than that of pure PP, indicating that the crystallization rate of whisker-filled PP accelerates slightly at a slow cooling rate; but at faster cooling rates, the crystallization rates of pure PP and filled PP are basically the same. Furthermore, the n values of pure PP are between 3.26 and 4.68, and the n values of whisker-filled PP are between 4.42 and 5.41, which are larger than that of pure PP, suggesting that the nucleation and

Table 6.7 Nonisothermal Crystallization Kinetic Parameters of Pure PP and 5% Filled PP

$\Phi(°C/$ minute)	Pure PP				5% Filled PP			
	$t_{1/2}$ (minutes)	N	Z	Z_C	$t_{1/2}$ (min)	n	Z	Z_C
5	1.61	4.68	0.07	0.60	1.51	4.83	0.10	0.63
10	0.88	4.51	1.21	1.02	0.87	4.67	1.36	1.03
15	0.69	3.12	2.23	1.05	0.66	5.41	6.60	1.13
20	0.45	3.26	9.10	1.12	0.47	4.42	19.83	1.16

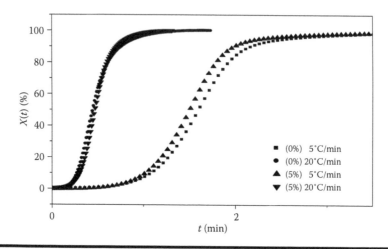

Figure 6.29 $X(t) - t$ **curves of two materials at cooling rates of 5°C/minute and 20°C/minute.**

growth of whisker-filled PP are complicated, and the three-dimensional growth of PP spherocrystals is probably hindered by fibrous whiskers; in addition, by comparing the Z_C values of two materials at different cooling rates we can see that the crystallization rate of whisker-filled PP is slightly higher than that of pure PP.

Figure 6.29 shows the crystallization curves of pure PP and 5% whisker-filled PP at cooling rates of 5°C/minute and 20°C/minute, respectively.

As can be seen from Figure 6.29, at the cooling rate of 5°C/minute, the crystallization rate of whisker-filled PP is slightly higher than that of pure PP, whereas at the cooling rate of 20°C/minute, the crystallization rates of two materials are almost the same.

6.4.2.2 Thermal Stability Analysis

The thermogravimetric curves of pure PP and PP filled with different amounts of calcium carbonate whiskers are shown in Figure 6.30, and the thermal decomposition temperature data at 5% thermal weight loss are listed in Table 6.8.

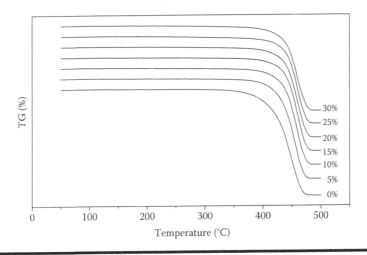

Figure 6.30 Thermogravimetric curves of pure PP and filled PP.

Table 6.8 Thermal Decomposition Temperatures of PP before and after Filling

Filling Content (%)	0	5	10	15	20	25	30
T_d (°C)	376.8	399.5	406.0	409.5	409.2	409.0	407.2

Note: T_d is the corresponding temperature at 5% thermal weight loss.

As can be seen from Table 6.8, at 5% thermal weight loss, the thermal decomposition temperatures of composites increase significantly with increasing filling content of calcium carbonate whiskers. The largest thermal decomposition temperature reached is 409.5°C at 15% filling content, indicating that the thermal decomposition temperature of PP material increases and thermal stability is improved with the addition of calcium carbonate whiskers.

6.4.3 *Mechanical Properties of Composite Materials*

6.4.3.1 *Analysis of Tensile Properties*

The measured tensile properties of composite materials are listed in Table 6.9.

Table 6.9 Tensile Properties of Composite Materials

Filling Content (%)	0	5	10	15	20	25	30
Tensile stress (MPa)	15.3	18.2	16.8	16.5	16.3	15.2	14.5
Tensile strain at break (%)	13.5	17.3	16.9	16.9	17.5	15.6	14.0

As can be seen from Table 6.9, the tensile stress of PP first increases and then decreases with increasing filling content of calcium carbonate whiskers, and reaches the largest value of 18.2 MPa at 5% filling content, an increase of 18.8% over pure PP. The tensile fracture strains of composite materials are all higher than that of pure PP, and reach the largest value, 17.5%, at 20% filling content, an improvement of 29.9% over that of pure PP, indicating that calcium carbonate whiskers have both enhancing and toughening effects on PP. The toughening effect especially is more significant.

6.4.3.2 Bending Performance Analysis

The bending stress and bending fracture strain data of the composites before and after filling are listed in Table 6.10.

From Table 6.10 we can see that the bending stress of composite materials increases when filled with 5%–20% calcium carbonate whiskers; the bending stress of the composite material filled with 5% whiskers increases 23.4% over that of pure PP; the fracture bending strain of the composite materials filled with 20%–30% whiskers increases significantly.

Table 6.10 Bending Performance Data of the Composites before and after Filling

Filling Content (%)	0	5	10	15	20	25	30
Bending stress (MPa)	29.9	36.9	34.6	34.9	33.3	33.1	29.7
Bending fracture strain (%)	9.0	10.05	7.6	9.0	15.1	14.7	15.9

6.4.4 Microstructure Analysis of Composite Materials

The tensile cross section and bending section of composites are observed using SEM, and the SEM images of tensile fracture sections are shown in Figures 6.31 and 6.32.

As can be seen from Figure 6.31, the fracture mode of the composite material filled with 5% whiskers is a mixed brittle and ductile fracture; and the fracture of 15% whisker-filled composite material is a brittle fracture.

In Figure 6.32a, the amplified tensile fracture section image of 5% filled material, we can see tensile deformation of calcium carbonate whiskers and the holes out of that part of the

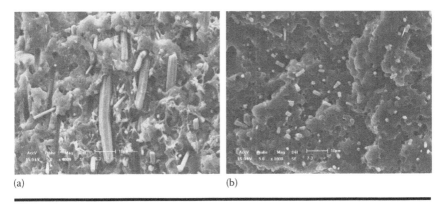

(a)   (b)

Figure 6.31 SEM images of tensile fracture sections (×500). (a) 5% whisker content. (b) 15% whisker content.

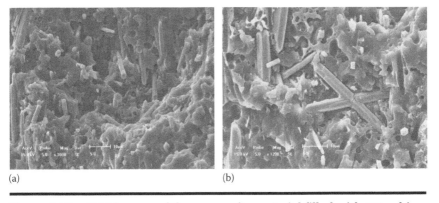

(a)   (b)

Figure 6.32 SEM images of the composite material filled with 5% whiskers. (a) 5% whisker content (×1000). (b) 5% whisker content (×1200).

whiskers are pulled. Figure 6.32b shows tetragonal calcium carbonate whiskers, which could be the result of a secondary growth of whiskers in the process of melt blending because the whiskers in the raw material are fibrous.

The SEM images of bending fracture sections are shown in Figures 6.33 and 6.34.

As can be seen from Figures 6.33 and 6.34, there are big and deep toughening nests and a large amount of broken calcium carbonate whiskers on the bending fracture section, indicating that the material produces a large plastic deformation during the breaking process. Whiskers in the matrix transfer and disperse stress through effective stress points and

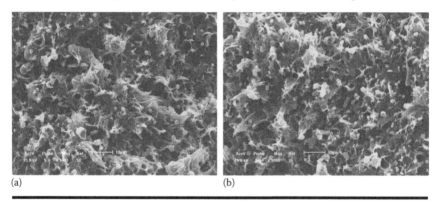

(a) (b)

Figure 6.33 SEM images of the bending fracture sections (×1000). (a) 5% whisker content. (b) 15% whisker content.

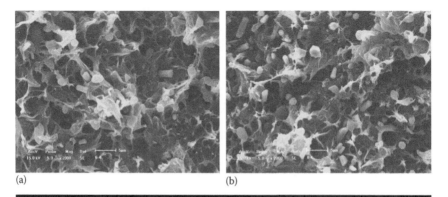

(a) (b)

Figure 6.34 SEM images of bending fracture sections (×2000). (a) 5% whisker content. (b) 15% whisker content.

consume more energy through its own breakage, and therefore achieve the goal of toughening.

6.5 Performance Analysis of the Composite Materials Prepared Using Extrusion and Injection Molding

6.5.1 Structure Analysis of Composite Materials

X-ray diffraction tests were conducted on composite materials,
and the results are shown in Figure 6.35.

As can be seen from Figure 6.35, the crystal form of PP is
still α type, and 2θ angles of five diffraction peaks are located
at 14.7°, 17.5°, 19.2°, and 22.0°, respectively, corresponding to
the diffraction produced by crystal face (110), (040), (130), and
overlapping (131). With the increasing amount of filled calcium
carbonate whiskers, the crystal form of PP does not change, but
the intensity of four diffraction peaks decreases gradually. In
addition, the diffraction peaks of calcium carbonate whiskers
are located at 36.9°, 39.1°, 43.6°, 46.3°, and 49.1°, and the diffraction peak intensity increases with an increasing amount of
whiskers, indicating that the calcium carbonate whiskers in the
composite material affect the crystallization properties of PP.

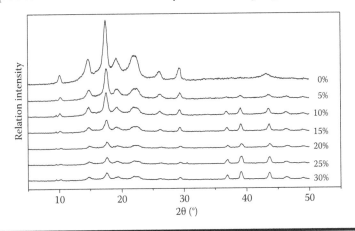

Figure 6.35   X-ray diffraction patterns of the composite materials.

6.5.2 Thermal Performance Analysis of Composite Materials

6.5.2.1 DSC Analysis

The DSC heating curves of pure PP and the PPs filled with different amounts of calcium carbonate whiskers are shown in Figure 6.36. The cooling curves are shown in Figure 6.37; both heating and cooling rates are 20°C/minute. The thermal analysis data are shown in Table 6.11.

As can be seen from Table 6.11, after filling with calcium carbonate whiskers, the melting temperature T_m of the composite material does not change significantly; the crystallization temperature quickly decreases first and then flattens after the whisker content is 5%. Both the melting enthalpy and crystallization enthalpy decrease with increasing whisker content. In addition, T_{co} decreases quickly with the increment of whisker content. The T_{co} of pure PP is 128.9°C, while the T_{co} of the PP filled with 5% whiskers decreases to 125.0°C; after that, T_{co} changes slightly with increasing whisker content, indicating that the crystallization temperature of PP decreases with the addition of calcium carbonate whiskers. The calculated ΔT results show that there is only a small change in overall ΔT. The ΔT of pure PP is 24.9°C, the ΔT of 5% filled PP decreases to 22.5°C,

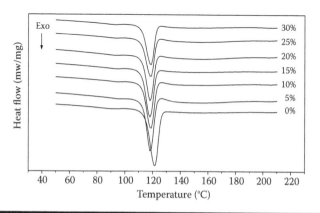

Figure 6.36 DSC heating curves of the PPs filled with different amounts of whiskers.

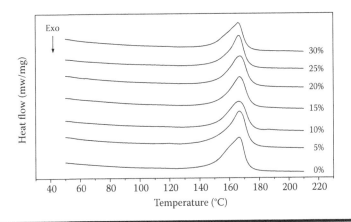

Figure 6.37   DSC cooling curves of the PPs filled with different amounts of whiskers.

Table 6.11   **Thermal Performance Analysis of PPs with Different Filling Contents**

Filling Content (%)	T_m (°C)	ΔH_m (J/g)	T_c (°C)	ΔH_c (J/g)	T_{co} (°C)	T_{cq} (°C)	ΔT (°C)
0	166.5	70.70	121.4	89.26	128.9	104.0	24.9
5	166.1	68.39	117.8	82.08	125.0	102.5	22.5
10	166.4	62.40	118.6	78.25	125.8	103.5	22.3
15	166.9	60.80	118.0	76.53	125.4	103.3	22.1
20	167.3	59.06	118.1	70.62	124.9	102.7	22.2
25	166.4	56.33	118.5	66.02	124.9	101.4	23.5
30	166.1	51.24	118.5	65.33	125.0	102.6	22.4

which essentially remains the same with an increasing amount of whiskers, indicating that the crystallization peaks of PPs slightly narrow after filling with calcium carbonate whiskers.

6.5.2.2 Thermal Stability Analysis

The thermogravimetric curves of PP composites filled with different amounts of calcium carbonate whiskers are shown in Figure 6.38, and the data are shown in Table 6.12.

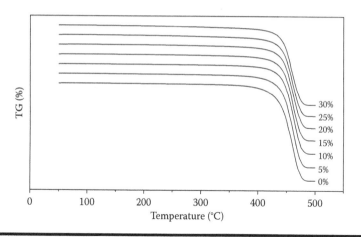

Figure 6.38 Thermogravimetric curves of pure PP and composite materials filled with whiskers.

Table 6.12 Thermogravimetric Data of Composite Materials

Filling Content (%)	0	5	10	15	20	25	30
T_d (°C)	405.5	413.3	421.0	420.1	419.6	416.5	415.8

Note: T_d is the corresponding temperature at 5% thermal weight loss.

As can be seen from Table 6.12, at 5% thermal weight loss, the thermal decomposition temperature of composite materials first increases and then decreases with an increasing amount of calcium carbonate whiskers. When the calcium carbonate whisker content is 10%, the thermal decomposition temperature reaches the maximum value of 421.0°C. This is because calcium carbonate whiskers have good thermal stability; the thermal decomposition temperature of the composite material filled with whiskers increases, and thus the thermal stability is improved.

6.5.3 Mechanical Properties of Composite Materials

6.5.3.1 Tensile Properties Analysis

The tensile stress–strain curves of pure PP and 10% whisker-filled composite material are shown in Figure 6.39. The tensile

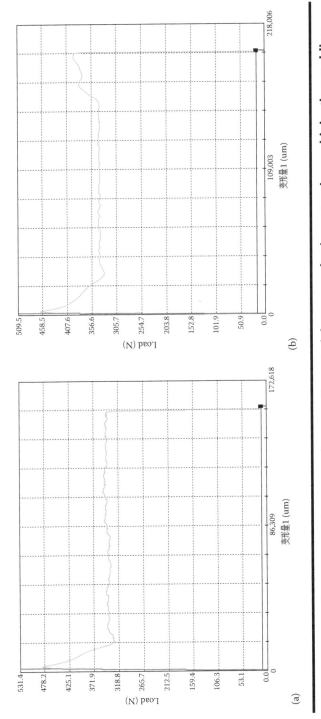

Figure 6.39 Tensile stress–strain curves of composite materials prepared using extrusion and injection molding. (a) Pure PP. (b) 10% whisker content.

curve of the composite material prepared using injection molding shows obvious yield points, and therefore belongs to the fracture with yield points. The tensile strength and tensile elongation at break data are shown in Table 6.13.

As can be seen from Table 6.13, compared with the yield stress and elongation at break of pure PP, with the addition of calcium carbonate whiskers, the yield stress of composite materials slightly decreases, but the elongation at break decreases significantly over that of pure PP. In addition, the breaking strength of the composites increases significantly with an increasing amount of calcium carbonate whiskers, indicating that orientation of both PP and whiskers occur after the yielding point, which improves the fracture strength of the composite material.

The tensile stress–strain curves of composite materials prepared using internal mixer-compression molding and extrusion molding are shown in Figures 6.40 and 6.41, respectively. Both tensile curves are fractures without a yielding point, but the elongation is obviously larger than that of pure PP, indicating that the toughness of composite materials increases after filling with whiskers, and the performances of composite materials prepared using different molding methods vary significantly.

Table 6.13   Tensile Properties Data of Composites

Filling Content (%)	0	5	10	15	20	25	30
Yield strength (MPa)	26.0	24.6	24.0	24.5	24.1	23.8	23.1
Breaking strength (MPa)	10.9	6.62	14.1	10.9	14.5	14.4	14.5
Elongation at the break (%)	393.14	340.21	321.47	201.83	232.20	212.89	209.01

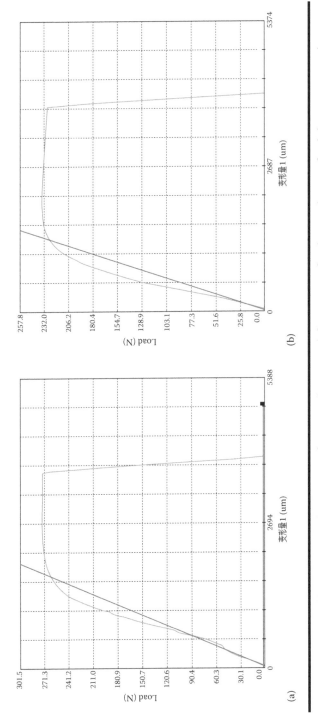

Figure 6.40 Tensile stress–strain curves of composite materials prepared using an internal mixing–compression molding method. (a) Pure polypropylene. (b) 10% whisker content.

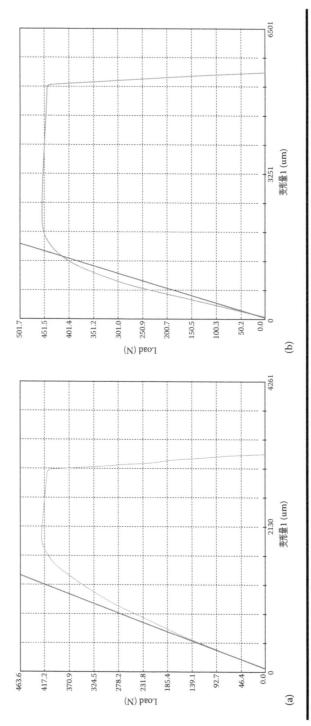

Figure 6.41   Tensile stress–strain curves of composite materials prepared using extrusion-compression molding. (a) Pure polypropylene. (b) 10% whisker content.

Table 6.14   Bending Strength and Bending Fracture Strain Data of Composite Materials

Filling Content (%)	0	5	10	15	20	25	30
Bending stress (MPa)	34.6	35.0	36.2	39.5	38.2	37.1	36.4
Bending fracture strain (%)	22.58	21.66	21.71	21.44	21.61	21.92	21.48

6.5.3.2 Bending Performance Analysis

The bending strength and bending fracture strain data of composite materials are listed in Table 6.14.

Table 6.14 shows that the bending stress of the composite materials filled with calcium carbonate whiskers increases significantly; the bending stress of the composite filled with 15% whiskers is the largest, up to 39.5 MPa, but the bending fracture strain decreases slightly, indicating that calcium carbonate whiskers have both reinforcing and toughening effects.

6.5.4 Microstructure Analysis of Composite Materials

The tensile fracture sections are observed using SEM, and the images are shown in Figure 6.42. Breaking fragments of

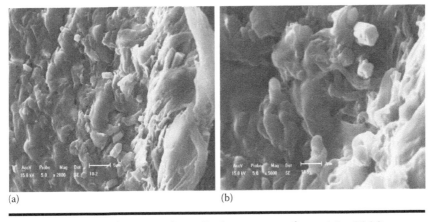

(a)   (b)

Figure 6.42   SEM images of tensile fracture sections. (a) ×2000. (b) ×5000.

whiskers and deformation of whiskers produced in the stretching process can be clearly seen in Figure 6.42, indicating that whiskers orient in response to tensile yield, which results in an increase of fracture stress.

6.6 Problems to Be Solved

1. The reinforcing and toughening mechanisms of whiskers need further study. Although surface treated inorganic whiskers can evenly disperse in polymers without anisotropy, and have reinforcing and toughening effects, their *L/D* ratios are still far smaller than those of glass fiber and carbon fiber, and their action mechanism is also different from that of glass fiber and carbon fiber. There are only a few studies on the reinforcing and toughening mechanics of whiskers filling polymers, so further study on the mechanism of action of whiskers can contribute to the development and application of whiskers.

2. The interfacial bonding of whiskers and matrix resin needs further improvement. Most polymers, such as polyethylene, PP, have very small polarity, while whiskers have very large polarity, although on surfaces treated with coupling agents, as can be seen from SEM images, interfacial bonding still needs further improvement. Interfacial bonding and interfacial interactions are key problems to be solved to obtain whisker-filled composite materials with better performance.

3. The preparation of composite materials usually needs two to three processes, such as the surface treatment of whiskers, melt blending, compression molding or injection molding, and so forth. Especially during the surface treatment process of whiskers and the blending process, whiskers are inevitably broken or damaged by strong mixing and high-speed stirring; thus the *L/D* ratios of whiskers are lowered and the reinforcing and toughening effects of whiskers are

limited. So, improving the toughness of whiskers and optimizing the material preparation process will be conducive for whiskers to achieve their full performance.

4. The relative higher price of whiskers than commonly used inorganic powder fillings is still the main restriction in their development. Therefore, reducing the production cost of whiskers by optimizing and improving their synthesis process will be beneficial to the application and development of whisker filling.

References

1. Jun Liu, Jiangping Li. Modification research progress of polypropylene. *Guangdong Chemical Industry*, 37(1):66–67, 2010.
2. Ting Liu, Xiaoying Cheng, Xuejun Zhang. Chemical modification of polypropylene. *Plastic Additives*, 13(6):20–23, 2010.
3. Haiping Wang, Biaobing Wang, Yunfeng Yang. Research progress in toughening modification of polypropylene. *Insulating Materials*, 42(1):29–32, 2009.
4. Heqing Fu. *Study on grafting modification of polypropylene and chlorinated polypropylene*. Guangzhou: South China University of Technology, 2005.
5. Yingjie Zhang, Jizhao Liang. Physical modification of polypropylene. *Shanghai Plastics*, 9(3):9–12, 2007.
6. Tongkao Xu. Production, application and market of filled plastics. *Plastics Processing*, 41(1):30–36, 2006.
7. Jianping Dong, Xianlan Ji, Yanwu Lei. Economic evaluation of energy-saving and reduce consumption in reconstruction project. *Chemical Industry*, 25(9):335–338, 2007.
8. Fengbo Cui, Guorong Cao. Research on properties of glass fiber reinforced polypropylene. *Fiber Glass*, 33(1):9–12, 2011.
9. Weixing Wang. *The research of nucleator and inorganic particles composite modifying PP*. Taiyuan: Taiyuan University of Technology, 2009.
10. Deqiang Zhang, Xuejia Ding, Wenjuan Chu et al. Phase behavior optical and mechanical properties of PP/SEBS binary blend. *Chemical Industry and Engineering Progress*, 29(7):1281–1286, 2010.

11. Bo Kong. *Study on liquid crystalline polymers as a new β-crystal nucleating agent to induce crystallization behavior of PP.* Shenyang: Northeastern University, 2009.

12. Meili Guo, Delu Zhao. *Polymer physics.* Beijing: Beijing University of Aeronautics and Astronautics Press, 2005.

13. Melvin Avrami. Kinetics of phase change. I. General theory. *Journal of Chemical Physics,* 7(12):1103–1112, 1939.

14. Andrzej Jeziorny. Parameters characterizing the kinetics of the non-isothermal crystallization of poly(ethylene terephthalate) determined by D.S.C. *Polymer,* 19(10):1142–1144, 1978.

15. Chao Zhao, Jian Zhou. Application of calcium sulfate whisker in HIPS. *Modern Plastics Processing and Application,* 21(4):45–48, 2009.

16. Qingfeng Liu, Desheng Wang, Shang Wenyu. Preparation of $CaCO_3$ whisker and characteristics of PP reinforced by whisker. *China Plastics Industry,* 28(1):5–7, 2000.

17. Wenyu Shang, Darong Xie, Qingfeng Liu et al. Properties of polymer material filled with aragonite whiskers. *China Plastics,* 14(3):24–27, 2000.

Index

Page numbers followed by f and t indicate figures and tables, respectively.